Exploring Representation in Evolutionary Level Design

Exploring Representation in Evolutionary Level Design

Synthesis Lectures on Games and Computational Intelligence

Editor
Daniel Ashlock, *University of Guelph*

Synthesis Lectures on Games & Computational Intelligence is an innovative resource consisting of 75-150 page books on topics pertaining to digital games, including game playing and game solving algorithms; game design techniques; artificial and computational intelligence techniques for game design, play, and analysis; classical game theory in a digital environment, and automatic content generation for games. The scope includes the topics relevant to conferences like IEEE-CIG, AAAI-AIIDE, DIGRA, and FDG conferences as well as the games special sessions of the WCCI and GECCO conferences.

Exploring Representation in Evolutionary Level Design
Daniel Ashlock

ISBN: 978-3-031-00992-1   paperback
ISBN: 978-3-031-02120-6   ebook
ISBN: 978-3-031-00169-7   hardcover

DOI 10.1007/978-3-031-02120-6

A Publication in the Springer series
SYNTHESIS LECTURES ON GAMES AND COMPUTATIONAL INTELLIGENCE

Lecture #3
Series Editor: Daniel Ashlock, *University of Guelph*
Series ISSN
Print 2573-6485   Electronic 2573-6493

Exploring Representation
in Evolutionary Level Design

Daniel Ashlock
University of Guelph

SYNTHESIS LECTURES ON GAMES AND COMPUTATIONAL INTELLIGENCE #3

ABSTRACT

Automatic content generation is the production of content for games, web pages, or other purposes by procedural means. Search-based automatic content generation employs search-based algorithms to accomplish automatic content generation. This book presents a number of different techniques for search-based automatic content generation where the search algorithm is an evolutionary algorithm. The chapters treat puzzle design, the creation of small maps or mazes, the use of L-systems and a generalization of L-system to create terrain maps, the use of cellular automata to create maps, and, finally, the decomposition of the design problem for large, complex maps culminating in the creation of a map for a fantasy game module with designer-supplied content and tactical features.

The evolutionary algorithms used for the different types of content are generic and similar, with the exception of the novel sparse initialization technique are presented in Chapter 2. The points where the content generation systems vary are in the design of their fitness functions and in the way the space of objects being searched is represented. A large variety of different fitness functions are designed and explained, and similarly radically different representations are applied to the design of digital objects all of which are, essentially, maps for use in games.

KEYWORDS

content generation, procedural content generation, evolutionary computation, level design, terrain maps, automatic design, cellular automata

Contents

Preface

This book summarizes six years of work with several collaborators on using evolutionary computation to design puzzles, mazes, level maps for games, height maps that specify terrain, and entire modules for use in a tabletop role-playing game. This books assumes at least a modest familiarity with evolutionary algorithms on the part of the reader.

The work started with the question of whether digital evolution could generate good test problems for robot navigation. After obtaining an affirmative answer and publishing a paper on the topic, I was very surprised when Julian Togellius, a fellow member of the IEEE Games Technical Committee, included the robot path planning work in a presentation surveying research on automatic level design for games. It was obvious to Julian that I had created a tool for automatically creating maps with an evolutionary algorithm. Returning home, I was faced with the problem of designing a death trap of the sort super villains use to torment superheroes. This problem arose from my weekly gaming group, not an academic setting.

These two events led to a whole series of papers written with Cameron McGuinness and Colin Lee, and later Laura Bickley, on creating various types of game content with evolution. The central issue was how to represent the material to permit the efficient evolution of a variety of interesting types of content. From a simple grid specifying full and empty space to a complex state conditioned version of a Lindenmeyer system and a well-chosen type of cellular automata that can crank out map after map, this work has been fascinating.

The normally vexing problem of creating a fitness criterion that permits evolution to effectively search for a solution, in this case a map, was not the difficult part of this work. Instead, the *same* fitness functions—one of a small set used in this book—could be applied again and again with different representations of maps or terrain, yielding very different classes of solution. Representation has been a central topic in my research across evolutionary computation and this is visible and prominent in the work presented in this book on automatic content generation.

Daniel Ashlock
April 2018

Acknowledgments

The work of my collaborators Colin Lee and Cameron McGuinness has been invaluable to the creation of this book. Cameron, especially, has been a collaborator on all the work except for the material in Chapter 4. Laura Bickley rode shotgun with me on the work in Chapter 4, on cellular automata. My wife Wendy, who normally collaborates with me on biology and mathematical game theory, deserves special thanks for serving as a proofreader on this text. I would also like to thank the dozens of students and friends who have been part of my gaming group over the years, setting the stage for much of the work.

Daniel Ashlock
April 2018

CHAPTER 1

Introduction

Level generation for games, particularly the problem of generating maps for those levels, is an area within the field of *Procedural Content Generation* (PCG). The goal of PCG is to provide a map for a level in a game together with populating that level with other game content. This book will examine the issue of *representation* in the context of using evolutionary algorithms [6] for PCG. A key point is that changing the representation, even while retaining many of the other algorithmic details, can radically change the type of content that is generated. The evolution of maps is an example of *search-based procedural content generation* (SBPCG), a variant of PCG that incorporates search rather than generating acceptable content in a single pass. SBPCG is typically used when a single pass will not suffice to locate content with the desired qualities. A survey and the beginnings of a taxonomy of SBPCG can be found in [69].

Although automated level generation in video games can arguably be traced back to the roguelikes of the 1980s (Rogue, Hack, NetHack, etc.), the task has recently received some interest from the research community. In [65] levels for 2D sidescroller and top-down 2D adventure games are automatically generated using a two-population feasible/infeasible evolutionary algorithm. In [70] multiobjective optimization is applied to the task of search-based procedural content generation for real time strategy maps. In [42] cellular automata are used to generate, in real time, cave-like levels for use in a roguelike adventure game, something that will be revisited and extended in Chapter 4.

An example of a map created by hybridization of two evolved structures appears in Figure 1.1. This map represents an underground dungeon. It is made of 16 tiles, each produced with an evolutionary algorithm and a simple evolved structure that specifies the connectivity that the tiles must possess is used to make a coherent map. The evolution needed to make this map required about six hours of computer time, but each tile is drawn from an evolved collection of tactically equivalent tiles meaning that once the computational groundwork is completed, an astronomical number of such maps can be created on demand in real time.

1.1 EVOLUTIONARY COMPUTATION

Evolutionary computation is any of a variety of optimization and modeling algorithms inspired by Darwin's theory of evolution. Since this theory is complex as it appears in nature, and is not completely understood at present, algorithms derived from it are more or less cartoon copies of the biological process. The simplest form of an evolutionary algorithm uses the steps in Fig-

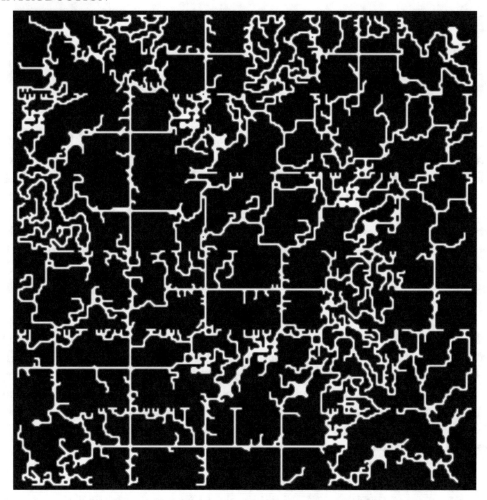

Figure 1.1: An example of a map generated by PCG techniques from Chapter 5.

ure 1.2. This basic algorithm can be elaborated in a number of ways, as we will see in later chapters.

Much of the research in evolutionary computation is concerned with the details of each of the steps shown in Figure 1.2. The definitions of the terms *mutation* and *crossover* are left intentionally vague at this point. Suffice to say that crossover mixes elements from two members of an evolving population, while mutation makes changes to the details of one member of an evolving population. The actual implementation of these operations is quite specific to both the problem being solved and the algorithm's encoding of that problem. Crossover plays the role of *sex* in biological evolution, while *mutation* is named after the biological process that inspires it.

Obtain an initial population of solutions
Evaluate the quality of the population members
Repeat
    Select population members with a quality bias
    Copy the selected population members
    Blend the copies via crossover
    Modify the copies with mutation
    Evaluate the quality of the modified copies
    Conditionally insert the copies into the population
Until(Done)

Figure 1.2: The steps of a minimal evolutionary algorithm.

The principle focus in this text is not associated with the standard details of evolutionary computation; rather it treats the issue of how the solutions making up the evolving population are encoded. The encoding is necessarily closely linked to the way crossover and mutation are done. This is the *representation* issue. While the details of the evolutionary algorithm are important, they typically represent no more than an available 100% improvement in performance. Modification of the representation can speed an evolutionary optimizer by amounts as high as a billion-fold [11, 22] and, in some cases, make it possible to solve the problem at all [24].

Evolutionary algorithms are *population-based* methods of solving problems. An evolutionary algorithm operates on a collection of solutions, called a *population* in imitation of biological populations. Almost every problem in this book will be phrased as an optimization against some quality criterion—called a fitness function in the parlance of evolutionary computation—and so the appropriate domain for discussion of *why* PCG is done with evolutionary computation is that of comparison to other optimization techniques.

One of the most widespread tools in optimization is the *gradient*. This is the vector of partial derivatives of the function being optimized. Imagine that the function being used as a quality measure for a population of maps like the one shown in Figure 1.1 is the ratio of distances between designated points within the maze with a very bad quality level awarded if those designated points are in the black regions (rock) as opposed to the white regions (open space). The independent variables are binary inputs, grid square by grid square, specifying if a given grid is open or full of rock. This means that there are no partial derivatives of any kind to be computed for the function.

This is one of the major benefits of using a population. Comparing the quality of different members of a population is, in effect, a discrete approximation of a gradient that points the way toward better structures in a high-dimensional discrete optimization space. This is not the usual view, rather it is one contrived to permit comparison with gradient-based optimization. The usual view is that we are using quality biased reproduction and selection in the presence of variation to imitate the powerful and successful process of biological evolution.

This viewpoint, that we are using an algorithm based on natural selection, raises some other issues. The initial population the algorithm starts with is one source of variation, but unless some clever heuristic is used in this initialization, this variation will mostly be among very low-quality structures. For evolution to follow its approximation to the quality or fitness gradient, there must be a good deal of variation. The use of *mutation* ensures a constant supply of variation to drive progress.

The key to the success of evolution, which quite literally works with random structures and random variation of those structures, is *selection*. The randomness is filtered and only good variations are retained. Good ideas can spread rapidly through reproduction while the roll of *crossover* is to mix and match ideas stored in different population members, juxtaposing them in the offspring those members have.

The selection method most commonly used in the remainder of the text is *single tournament selection* which also serves to place new structures back into the population. To perform single tournament selection of size n, a group of n structures is chosen from the population. The two most fit members of the group are selected to reproduce and their children replace the two least fit members. The selection method *tournament selection* picks two groups and the parents are the best in each or the groups, individually. Each method is able to simulate the other by changing the value of n and single tournament selection has slightly lower computational expense.

Taken together, mutation and crossover are called the *variation operators* that drive the search of a basic evolutionary algorithm. An active research topic is the creation of other variation operators as well as exploring the details of how and how often crossover and mutation are performed. An important part of characterizing an evolutionary algorithm is to check if crossover should be used at all, throughout evolution, or perhaps only during the initial stages. There are problems where each of these three strategies is the correct choice.

1.2   ELEMENTS OF FITNESS FOR LEVEL DESIGN

In order to use evolution as an optimizer, we need to be able to tell better structures from worse structures. This means we need a measure of quality that can be used to make comparisons. From this point on, quality measures for members of an evolving population will be referred to as *fitness functions*. When specifying the properties required in a level map, distances between specified points on the map are the basic elements from which fitness functions can be derived, although Chapter 3 will give a very different type of fitness function for *terrain maps*. Suppose that a map is always something that happens on a grid and that there is always a way of telling if you can move from one grid square to another or not. In this situation a type of algorithm called *dynamic programming* [34] is the natural tool.

Dynamic programming [30] is a ubiquitously useful algorithm. It can be applied to align biological sequences [55], to find the most likely sequence of states in a hidden Markov model that explain an observation [71], or to determine if a word can be generated from a given context free grammar [46]. The version of dynamic programming used in this study works by traversing a network while recording, at each network node, the distance traversed to arrive at that node. When the length of a new path is not superior to that already found, the search is pruned; otherwise, the minimum cost of reaching the node is updated and search continues. Dynamic programming has extraordinarily broad application, and so we will now explain how it is used to find distances in levels. A more detailed explanation of the use of dynamic programming in games appears in [19].

One of the earliest and simplest dynamic programming algorithms is *Dijkstra's algorithm*, which is used to sort out shortest paths in a combinatorial graph. For our purposes we view a level map as a graph in which the grid squares of the map are vertices and two vertices are connected if they represent grid squares where motion from one to the other is possible. The algorithm starts at a specified vertex of a graph, corresponding to an important location in the level, and works outward. Every edge not previously traversed, with one end at a vertex currently being traversed, is visited. The total cost of reaching the vertex (distance to the vertex) at the other end is then recorded—or possibly updated if the vertex has been visited before but a cheaper path has been found. Dijkstra's algorithm is not the best algorithm for finding a complete set of distances from a vertex [51], but it is much simpler to implement than the more efficient algorithms.

The version of the algorithm most often used in games finds distances on a game board or minimum number of moves needed to achieve an objective. The left image in Figure 1.3 shows how this plays out on a board that consists of a simple, unobstructed grid of squares. The vertices of the graph are the squares, and two squares are adjacent if they share a face. The starting point for Dijkstra's algorithm is the square marked with a zero. In this version of the algorithm— since the cost of all edges in the graph is equal—the first time you reach a vertex (square) you discover its actual distance from the starting point. While the use of an open square yields a very simple pattern of distances, Dijkstra's algorithm applies equally well to spaces with more complex connectivity. An example of this is shown in the right image in Figure 1.3.

In general, to find the distance from one point to another, Dijkstra's algorithm is run starting at one of the points, and the value at another is then saved. There are more efficient versions of the algorithm. A heuristic that chooses which grid square to examine next transforms Dijkstra's algorithm into one of the A^* family of algorithms, depending on the heuristic. The use of heuristics by the A^* algorithm can substantially speed the computation of distances and is frequently used in real-time games [57]. From this point on, we will assume that we can simply compute distances from any point in any grid-based map.

1.3   OBSCURED MAZES: A SIMPLE EXAMPLE

One of the simplest possible fitness functions that can be defined in terms of distance is to make the shortest path from start to finish as long as possible. One of the least complex representations for evolving level maps is the *binary* representation which stores one bit of information for each grid square on a level: full or empty. The level map in Figure 1.4 is an example of such an evolved

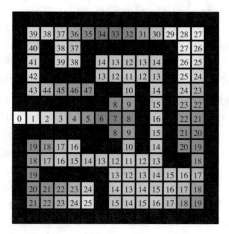

Figure 1.3: The computation of distances via Dijkstra's algorithm on an open grid of spaces (left) and in a more maze-like map (right). The starting point in both maps is the square with a distance of zero.

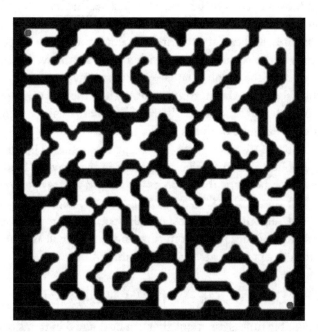

Figure 1.4: This maze was evolved with a simple binary representation—each grid is empty or full—with the goal of maximizing the distance between the two red dots.

map using both a binary representation and a longest path fitness function. The simple fitness function permits testing of other features of an evolutionary algorithm. Examining this maze, there is an obvious problem: this maze is very easy. At any point there is only a way forward and a way back. The fitness function that makes the distance from start to finish as long as possible will put all the available space into the path from start to finish to lengthen it. This means that culs-de-sac, blind alleys, and closed loops will all be avoided by the search algorithm as all this space is needed to maximize the nominal path length.

One might conclude that longest-path mazes are simple debugging tools, not really useful as actual content. It turns out that the simplicity of these mazes is a side effect of being able to see the whole maze or its connectivity within a region. If you do not have a top-down board or universal view then the maze can become more challenging. Two examples of *obscured mazes* that are instances of this thought are *chess mazes* and *chromatic mazes*.

1.3.1   CHESS MAZES

The author is part of a gaming group that meets once a week. At one point, during a game where the characters were members of a superhero team, a villain named *Mr. Unreasonable* had blackmailed the heroes into testing the death traps that he offered for sale to other villains. One of these was a maze made of chess pieces. Each chess piece was placed in the maze and, based on its moves, covered various squares in the maze. The characters had to traverse the maze, using the moves of a particular chess piece, without passing through any of the squares covered by the enemy chess pieces.

This is clearly a type of maze, defined by the placement of the enemy chess pieces, and it is an obscured maze because the "walls" are created by the rules of chess rather than explicitly displayed. An example appears on the left side of Figure 1.5. Notice that it is *hard* to see the squares the rook must not move through in the actual maze. The *key maze* colors the squares covered by knights blue and the ones that the rook may move through with red. A total of 18 rook moves are required to complete the maze. Recall that the rook must not pass through squares controlled by any of the enemy pieces.

After resolving the gaming session, it became obvious that an evolutionary algorithm using a dynamic programming based fitness function could be used to design chess mazes. The player has a piece on the board and other pieces, of the other color, are placed on the board to form the maze. A target position is given that the player's piece must be able to reach. The requirements are as follows.

1. It must be possible for the player's piece to reach the target square.

2. The number of moves required by the player to reach the target square must be maximized.

Notice that these rules destroy the original notion that two grid squares are adjacent, for dynamic programming purposes, if they share a face. In a chess maze two grid squares are adjacent if the player's piece can move between them.

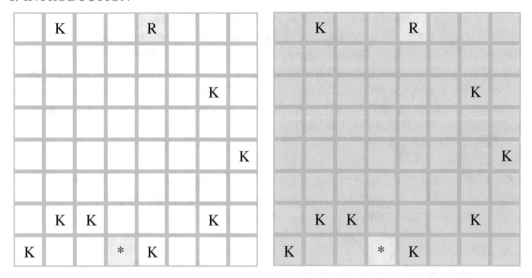

Figure 1.5: This is an example of a chess maze. The left version is the actual chess maze, the right version is the key. The player's piece is the rook (R) at the top of the board. The maze is defined by eight enemy knights and the goal square is the (*) at the bottom of the board.

The representation for the chess maze is simple. A maze is defined by the player's piece type, the start and goal squares, and the identity of the opponent chess pieces. Everything except the positions of the opponent's chess pieces are specified in advance. An order is fixed for the opponent's chess pieces to be placed on the board (for "eight knights" this is trivial, but we will have an example of a mixed opponent type chess maze later). The evolutionary algorithm operates on a list of grid square positions for the opponent's pieces. The constraints that each opponent piece must occupy its own grid square and not block the square where the player's piece is positioned initially are handled by awarding fitness zero to any set of coordinates that violate these constraints. Similarly, impassable mazes are awarded a fitness of zero. The issue of impassable mazes can become a serious one, something that is dealt with in Chapter 2 in a number of ways. Details of the evolution of chess mazes are given in [7].

After a search for past art, relatively few similar efforts were located. The chess mazes presented in this study, aside from not using a standard chess board, are very similar to the chess mazes introduced by Bruce Albertson as a type of puzzle used to teach chess moves to novice players [3, 4], although these are designed by hand. Related uses of evolutionary computation and artificial intelligence include the following. Multi-objective evolutionary computation has been used to create opponents in games such as driving agents [2] and to design combinatorial games [31] with the additional objective of finding playable games. Artificial intelligence has been applied to the problem of balancing board games by modifying their rules dynamically [49] and to the problem of dynamically adjusting the difficulty of a video game [41]. In [68] the authors use an evolutionary algorithm to select rules for a game. None of these papers seeks to create a palette of game elements similar to those produced in this study.

A key point is that the longest path maze becomes interesting again when the "walls" of the maze are not easy to see. The key maze given in the right half of Figure 1.5 is a useful design tool. Not only does it make it easy for a human to validate the maze, but it also permits a designer to see the mazes character *without* needing to visualize all the places the opponent's chess men can move. Some additional examples of chess mazes, shown as keys, are in Figure 1.6.

The examples in Figure 1.6 speak to another important point. Evolutionary algorithms are stochastic optimizers. Their outcome, while often high quality, depends on random factors. There are times when this might be problematic, but in the case of ACG it is a boon. The algorithm looking for a chess maze with eight knights that can be solved by a player with a rook was run 100 times and discovered 100 distinct chess mazes. This leaves quite a bit of room for a puzzle designer, or someone looking for puzzles to put in a book to look for other aesthetic factors. The mazes shown thus far have fit on a standard 8 × 8 board but, in fact, the evolutionary algorithm has no problem finding larger boards. An example is shown in Figure 1.7.

The large example in Figure 1.7 is somewhat challenging and, inspecting the maze, culs-de-sac, and alternate routes make an appearance. The multiple diagonal paths from the upper right to the lower left are an example of a multiple route which must be departed in the middle of a segment to solve the maze. This is a good example of how representation affects the type of maps that evolve. In Figure 1.4, a maze is given in which evolution allocated so much of the available space to satisfying the long path fitness that the maze lacks deceptive features; the chess maze has them in spite of long path fitness precisely because the representation is too course-grained to permit efficient allocation of all space to the path. Placing 14 opponent chess pieces on 439 grid squares (all but the start and finish) does not permit efficient allocation of empty space.

A final tactical point in case a reader wishes to evolve their own chess mazes: when a chess maze has rooks, bishops, or queens in its set of opposing pieces, then pawns are needed to *prevent* those pieces from covering so much of the board that they make the problem of finding a traversable maze impossible. This phenomenon is highly active in the example in Figure 1.7.

1.3.2   CHROMATIC MAZES

Chess mazes are obscured by the fact that a human being, other than one that is very good at chess, cannot see the obstructions caused by the opposing chess pieces without a lot of effort. *Chromatic mazes*, implementing an alternate strategy, are made obscure through the use of obscure connectivity. The chromatic puzzle is created by assigning the colors **R**(ed), **O**(range), **Y**(ellow), **G**(reen), **B**(lue), and **V**(iolet) to the squares of a grid. Safe moves consist of stepping from a color to the same color or one adjacent to it on the color wheel. The difficulty of a puzzle is again determined by the number of moves required to traverse the puzzle. A secondary factor that can be computed is the number of safe moves that require a change of colors. The more

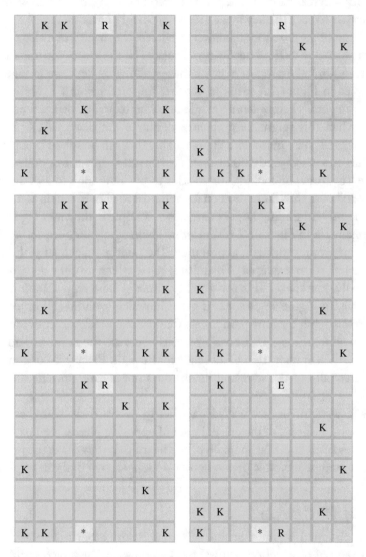

Figure 1.6: Six chess-maze keys for a rook player moving against eight knights. The fitness is the number of rook moves required to complete the maze; no mazes of fitness 17 were found in 100 runs of the evolutionary algorithm—and only one of fitness 13.

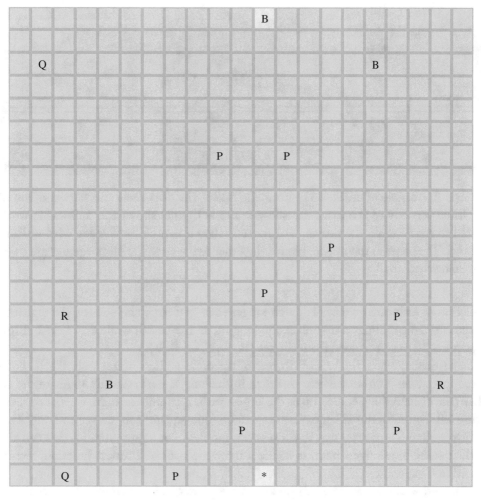

Figure 1.7: The key map for a 21 × 21 chess maze in which a bishop seeks to navigate a maze created with eight pawns, two bishops, two rooks, and two queens.

color changes that are required the less visible the maze is as a monochrome set of squares. We do not compute this factor here as the use of six colors, in combination with the puzzle evolution strategy, yields relatively few monochromatic multi-square fragments in the solutions to the evolved puzzles. Figure 1.8 gives an example of an evolved chromatic maze.

While chromatic mazes are a second example of a type of obscured maze, they are markedly different from chess mazes. A different type of mental effort is needed to see where the path lies in a chromatic maze. The representation for the chromatic maze is simply a list of the color values that fill the maze in reading order (row by row). This means that the "genetic"

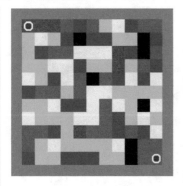 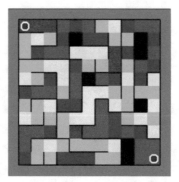 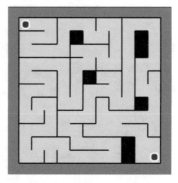

Figure 1.8: A example of an 11 × 11 chromatic maze. The dots show the start and end of the maze. Three versions of the maze are shown. The left panel shows the maze as it would be presented to a person do the puzzle. The maze is shown as a classical maze with walls in the right panel. The panel in the middle blends the other two views. The black squares are ones that cannot be reached at all.

specification of a chromatic maze is much longer than that for a chess maze. An eight-knight maze specifies 16 coordinates that place the knights on the board, while an 11 × 11 chromatic maze requires 121 values drawn from the set {**R, O, Y, G, B, V**}.

In particular, where the lack of degrees of freedom in chess mazes restored the potential for some space to end up somewhere else than in the longest path, the chromatic mazes have a very bad case of everything ending up in the largest path. Examine the maze view of the example in Figure 1.8. Only 5 of the 121 squares are spent on (one-square) culs-de-sac and only 7 end up being inaccessible. The correct way to address this problem, for representations that can allocate all of their space to a longest path, is to use a more sophisticated fitness function, something addressed in Chapter 2.

Comparing chess mazes and chromatic mazes gives us a first demonstration of the power of choice of representation. The two representations reacted very differently to the longest path fitness function *even though* the mazes were being generated by essentially the same evolutionary algorithm using the same fitness function. The two representations also share the same scalability. An example of a relatively large 21 × 21 chromatic maze is given in Figure 1.9. This maze does not allocate all the available space to the single path, but this is because the algorithm was not permitted additional time to evolve, in comparison to the evolution of 11 × 11 puzzles.

1.3.3   KEY MAPS AND ALTERNATE VIEWS

If automatically generated content is to be used, we face a problem for the human designer allocating and selecting that content. Examine the chromatic maze in Figure 1.9. It is incomprehensible at first glance. A big part of the solution is the use of *key maps* and other alternate views of the ACG. In Figure 1.8 there is a view that shows the chromatic maze as a classical

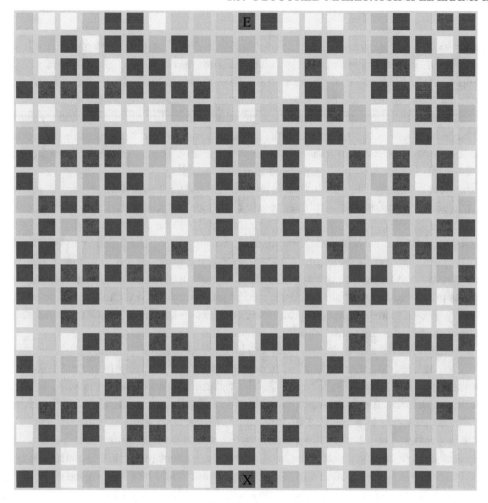

Figure 1.9: A 21 × 21 chromatic maze. The (E)ntrance and e(X)it are designated by letters. The maze requires 262 moves to complete on a board of 441 squares.

maze with walls. The initial example of a chess maze in Figure 1.5 shows another type of key map.

The key maps shown in this section are just examples; actually showing a solved version of the maze with a visible path would be another piece of useful information. When working with SBPCG, it is easy for the system to satisfy every goal set by the designer multiple times and still present an incomprehensible mess to the designer. This means that additional views, like but not limited to key maps, are potentially critical for the useability and adoption of an automatic content generation and management system.

1.4   CONCLUSIONS

The examples in this chapter apply a relatively straightforward evolutionary algorithm to generate examples of different types of puzzles: chess mazes and chromatic puzzles. The puzzles are all examples of different types of mazes that enhance problem difficulty by using *implicit barriers*, ones that must be deduced by the agent traversing the maze. These puzzles are intended to supply content in the form of mini-games or puzzles. The ability of an evolutionary algorithm to find large numbers of optima of a complex function means that the evolutionary algorithms presented here can generate libraries of puzzles very quickly. The assortment of small chromatic mazes shown in Figure 1.10 were generated in a few seconds.

The details of how the puzzles are integrated into a game are a matter for the game designer. If a player is unfamiliar with chess moves, or does not know that the color wheel is the critical clue designating safe moves, then they may make some missteps. If making an unsafe move is not deadly but merely does some damage to the player, the puzzle can be softened so that a player can discover the rules of the maze by experimenting. It is also not clear that a game designer would want to use the two types of puzzles presented here. This is another reason that the puzzles presented here should be considered as an example of a technique. They can be applied to any type of grid-based puzzle that the designer can specify a representation for. Such a representation specifies the contents of grid squares and the adjacency.

The technique developed in this study can be applied to implicit mazes of almost any type. It could even be used to generate level or structure layouts by adapting techniques like those in [17], something that appears in Chapter 2. The technique also could be applied to far more challenging structures. As an example, consider the following generalization of the chromatic maze. The actual maze is generated in three dimensions with some number of levels. These levels are not realized as *physical* dimensions: rather, when a player enters a red square, all the floor squares change color to match the level below the current one; a blue square will take the player up a level. Making sure such a maze is even solvable could be a substantial chore for hand design, even with a good digital design framework. An evolutionary algorithm could construct hundreds of such mazes, with certificates of solubility, per hour.

The fitness values obtained for the simple make-the-path-long puzzles were startlingly high. Since the fitness values are minimum number of moves required to safely traverse a puzzle, this leads to an undesired simplicity of the high-fitness puzzles; often there is only one choice of a way forward. Unless the player is not, initially, aware of what makes a safe move this means that the puzzles generated are rather simple and possibly tedious. This problem can be solved by the application of alternate fitness functions. Shifting to multi-criterion optimization that also rewards branching factor in the implied maze or other difficulty or playability factors is an early priority for future research.

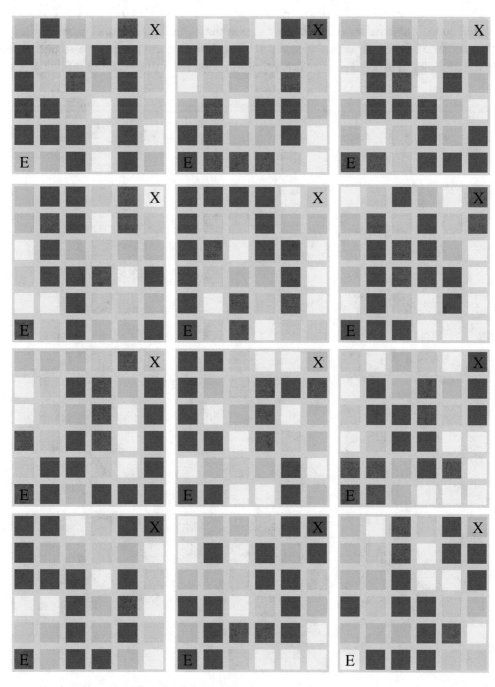

Figure 1.10: A collection of evolved 6 × 6 chromatic mazes.

Dynamic programming is not a terribly "human" way of solving a maze. It is both exhaustive and memory intensive. This suggests that a maze that is hard for dynamic programming may not be hard for a human player and vice versa. Using humaniform agents in a second phase to grade the difficulty of evolved puzzles is another direction for future research on this project.

CHAPTER 2

Contrasting Representations for Maze Generation

The material presented in this chapter grows out of the material presented in Chapter 1, where two sorts of mazes were generated. Both of these types of mazes were made more difficult for a player by the fact that the walls were *implicit* or not directly visible. In this chapter we will work with explicit, clearly visible mazes and explore the design of fitness functions and the impact of varying the representation that specifies the open and blocked portions of the mazes. The material in this chapter is an expanded version of material presented in [14].

The fitness function used in the earlier research computed the length of the shortest path from entrance to exit of the maze. When the fitness function was maximized, this had the effect of making the maze as long as possible, given the representation used to encode the maze. Maxima of this fitness function are winding paths that do not branch except by accident. The resulting evolved mazes have a feature that would be undesirable in many circumstances: the mazes generated are trivial, with the only remaining challenge being to deduce the rules for where it is possible to move. An alternate fitness function, which attempted to minimize the difference between the shortest path from entrance to exit and some targeted value, was introduced in [7] and resulted in mazes which contained some non-trivial features such as branching and culs-de-sac. These features, however, appeared more or less at random, suggesting the need for more sophisticated fitness functions.

The primary novel areas presented in this chapter are, first, to explore a variety of representations for encoding level maps and, second, to give a framework for designing fitness functions that give the user substantial control over the character of the mazes that are evolved.

Representation is a key issue in the design of an evolutionary algorithm [59]. Representation is a point at which domain knowledge can be incorporated. Four representations are presented in this chapter, and the primary effect of changing representation is not an improvement in algorithm efficiency but a change in the character of the maps located by evolution. These representations are as follows.

1. A *binary, direct* representation, in which open and blocked squares within a rectangular grid are specified directly as a long, binary gene with the values *full* and *empty*.

2. A *direct* representation, in which the squares within a grid are assigned colors from the set {red, orange, yellow, green, blue, violet}. These colors are specified directly as a long gene

over the alphabet {R, O, Y, G, B, V}. An agent can move between adjacent squares if they are (i) the same color or (ii) adjacent in the above ordering. This representation is the one used for chromatic mazes in Chapter 1, but we will use more complex fitness functions.

3. An *indirect positive* representation in which the chromosome specifies structures that are placed on an empty grid to form the level. In this representation, walls are explicit and rooms and corridors implicit.

4. An *indirect negative* representation in which the chromosome specifies material to remove from a filled grid to form the level. In this representation, rooms and corridors are explicit, while walls and barriers are implicit.

In each of these representations the issue of feasibility (can a player traverse the maze at all) is of some concern. In all three non-chromatic representations, the expressed maze amounts to a specification of obstructed and unobstructed squares on a rectangular grid of width X and height Y. In three of the representations, a technique called *sparse initialization* was used. Sparse initialization places relatively few obstructed squares in the mazes in the initial population. This makes the mazes in the initial population easy-to-solve, low fitness mazes but leaves all members of the initial population feasible. Greater complexity and higher fitness are then discovered incrementally by the variation operators over the course of evolution. The details of sparse initialization are given in the descriptions of each representation.

The two indirect representations are examples of *generative representation*. Generative representations are a fairly old idea. One of the early names for generative representations was *cellular encodings*. In [38, 72] the authors explore a cellular (generative) encoding for artificial neural nets (ANNS). This example is included because it is one of the earliest examples of a generative representation. *Evolutionary design* is the term for using evolutionary computation to perform design, and generative representations play a substantial role in this field. An early survey of evolutionary design appears in [45], while a more recent collection of papers on evolutionary design, including generative representations, is [39]. In this collection the terms *developmental encoding* and *developmental representation* are used in describing generative algorithms. In evolutionary computation, a generative representation is one that does not directly encode a solution to the problem of interest. Rather, it gives directions for constructing a solution.

Generative representations are used for a wide variety of tasks. In addition to the level design applications in this and later chapters, generative representations are used to search the space of networks. A collection of existing representations for network induction are presented in [23], although many lack generality and adaptability. The idea of parameterized a generative representation for graph evolution was first presented in [23], where it was found that the parameter settings had a significant impact on algorithm performance. The generative TADS representation for network induction was developed in [29]. TADS stands for Toggle, Add, Delete, and Swap. These are four commands that toggle an edge in a graph, add one, delete one, or take two edges that for two sides of a square in the graph and move them to be the diagonals of the graph. Toggling an edge adds it if it is absent and removes it if it is present. This representation uses a starting graph and has chromosome consisting of edits (toggle, add, delete, swap) to that starting graph. A general type of adaptive, generative representation is proposed and tested in [53].

Generative representations permit the incorporation of domain-specific information into an evolutionary computation system. The positive and negative indirect representation in this chapter, for example, directly incorporate the idea of "wall" and "room or passage." This means that these higher level structures do not need to be self-organized out of individual bits specifying if grid squares are open or full—which in turn enormously reduces the size of the space that the ACG system is searching. This is good because reducing the size of the search space pays bonuses in speed of search. It is *potentially* a bad thing because the restriction may exclude desirable solutions. This is not inevitable and speaks to the need for careful, thoughtful design of generative representations.

2.1 DETAILS OF THE BINARY DIRECT REPRESENTATION

For an $X \times Y$ board the direct representation is a string of XY bits. Two variation operators are used: uniform crossover with probability p_c and uniform mutation with probability p_m. The uniform crossover operates on two chromosomes, exchanging their bits at each location with independent probability p_c. Uniform mutation operates by flipping the bit at each location with probability p_m. Actual rates for these operators are given in the description of experiments in Section 2.6. In addition, two experiments are performed using traditional one- and two-point crossover. *One-point crossover* exchanges suffixes of genes while *two-point crossover* exchanges middle segments. An example of a maze evolved with the binary direct representation is shown in Figure 2.1.

The probability that a maze in which obstructed and unobstructed squares are equally likely can be traversed is so close to zero that at least one 500-member initial population, encountered during algorithm development, contained zero reversible mazes. In order to address this problem, a parameter $0 < fill < 1$ was added to the algorithm. It is the probability a given square will be obstructed (that a bit in the chromosome will be one). This parameter is set to $fill = 0.05$ in all experiments—a 1/20th full set of initial mazes. The effect of this is to start with traversable mazes of relatively low fitness and permit the variation operators to fill in added obstructions so as to increase fitness in a selection guided manner. We call this technique *sparse initialization.*

Given the history of evolutionary computation, the direct binary representation is a natural one. It also has the ability to generate uniformly infeasible initial populations. Without some thought and analysis, this representation is stillborn. The reasoning that leads to the idea of using sparse initialization is not sophisticated or difficult, but if the work were attempted with a generic tool using a standard representation, a researcher might arrive at the point of thinking

Figure 2.1: An example of an evolved maze using the binary direct representation. The entrance and exit are marked with large circles, smaller circles indicate checkpoints used by the fitness function.

the problem is evolution intractable (it would in fact be generic tool intractable). This is another point where discussion of representation is valuable—a problem intractable in one representation may be simple in another representation that is not even too different.

2.2   DETAILS OF THE CHROMATIC REPRESENTATION

For an $X \times Y$ board the second direct representation is a string of XY values in the range $0 \leq x \leq 5$ mapping onto {R, O, Y, G, B, V}, as in Chapter 1. Two variation operators are used: uniform crossover with probability p_c and uniform mutation with probability p_m. The uniform crossover operates on two chromosomes, exchanging their colors at each location with independent probability p_c. Uniform mutation operates by generating a new color at each location with probability p_m. Actual rates for these operators are given in the description of experiments in Section 2.6. An example of a maze evolved with the chromatic representation is shown in Figure 2.2.

Similarly to the binary direct representation, a form of sparse initialization is needed with the chromatic representation. The form of sparse initialization used is to initialize the grid to

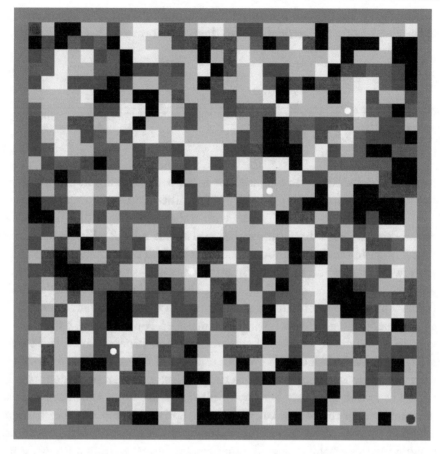

Figure 2.2: An example of a chromatic maze. The large black circles mark the entrance and exit; the smaller white dots mark checkpoints.

have squares that are green and yellow with those colors being equally likely. A maze initialized in this fashion permits movement between all pairs of squares while making it very easy for mutation to create barriers. The details of sparse initialization are quite different from those used in the binary direct representation, but the spirit—start with an easy, open maze and let evolution toughen it—remains the same.

This representation is very different from the others used in this chapter in the following sense. The other representations all specify which squares of a grid are obstructed. The chromatic representation does not directly obstruct any square within the grid: rather, it imposes rules about which squares it is possible to move between. This means that it creates implicit walls between squares that define the maze rather than defining it with obstructed squares. Implicit walls create a situation in which the number of bits in a minimal description of a chromatic maze is much larger, at the same grid size, than for the binary direct representation. This is because each *pair* of adjacent squares has a potential barrier between them rather than each square being a potential barrier.

2.3 DETAILS OF THE POSITIVE, INDIRECT REPRESENTATION

This representation is similar to the one used for robot path planning in [17]. It is stored as a linear array of 80 integers in the range $0 \le n \le 9999$. The integers are used in pairs to specify barriers in an originally empty $X \times Y$ arena with walls at its edges. The first integer is bit-sliced into a one-bit number, a three-bit number, and a remainder R which is taken modulo the larger of X or Y to yield the length of the barrier. The one-bit number determines if a barrier is penetrating or not. The three-bit number is interpreted as a direction for the barrier to run from its starting point W, NW, N, NE, E, SE, S, or SW. The second integer i is split as $x = i \ mod \ X$ and $y = ((i \ div \ X) \ mod \ Y)$ to obtain the starting point (x, y) of the barrier on the grid. The operator *div* represents integer division without remainder. A penetrating barrier runs from its starting point to its length or an edge of the arena. A non-penetrating barrier stops when it encounters any other barrier. Figure 2.3 gives an example of unpacking this representation.

There is an additional parameter used in the positive, indirect representation: it is permitted to lay down at most a fixed number *Lim* of filled squares. Once this number of squares is reached, the remainder of the barrier currently being laid down, and all other barriers following it, are skipped. Two variation operators are used: uniform crossover with probability p_c and uniform mutation with probability p_m. The uniform crossover operates on two chromosomes, exchanging integers at each location with independent probability p_c. Uniform mutation operates by replacing the integer at each location with a new integer chosen at random in the range $0 \le n \le 9999$ with probability p_m. Actual rates for these operators, the number of pairs of integers in a chromosome, and the maximum number of filled squares Lim are given in the description of experiments in Section 2.6. An example of a maze evolved with the positive representation is shown in Figure 2.4.

As with the direct representations, there is a very high probability that a maze created with this positive, generative representation will have no path from the entrance to the exit of the maze unless either the number of barriers is kept quite small or barriers start with very short lengths. A maze with few barriers is not likely to be interesting no matter how cleverly they are placed; for this reason initial barriers (but not those created by mutation) have lengths in the range $1 \le L \le 3$. This technique represents a third form of sparse initialization.

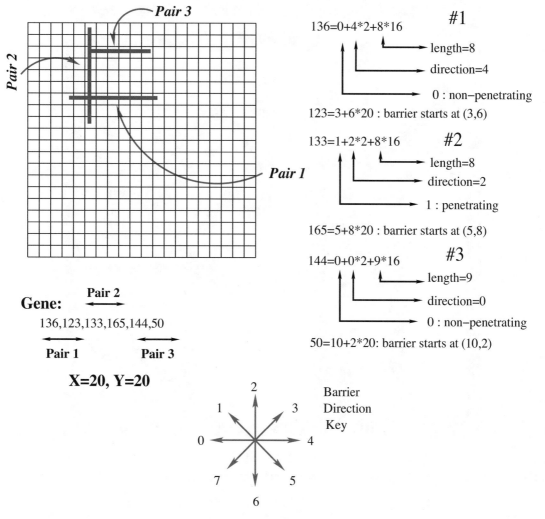

Figure 2.3: An example of unpacking a gene for the positive, indirect representation with six numbers (136, 123, 133, 165, 144, 50) specifying three barriers. The barriers are shown as thick lines running from their beginning grid square to their ending grid square. In practice, all squares containing a part of a barrier are registered as "full." Note that barrier #3 stops at barrier #2 after only expressing five of its nominal length of 9—this is an example of a non-penetrating barrier.

2.4 DETAILS OF THE NEGATIVE, INDIRECT REPRESENTATION

This representation is stored as a linear array of 120 integers in the range $0 \le n \le 9999$. The integers are used in pairs to decide where to remove material, in the form of rectangular rooms

Figure 2.4: An example of a positive, indirectly encoded evolved maze. The large red circles mark the entrance and exit while the smaller green ones mark checkpoints.

and corridors, from an initially filled $X \times Y$ grid. A pair of 4×4 rooms are always placed over the entrance and exit of the maze to increase their "size" in the dynamic programming environment, decreasing the chance a given specification of material removal will fail to join the entrance and exit.

The first integer in a pair is bit-sliced into a two-bit number and a remainder R which is used to determine corridor length or room dimensions. The two-bit number determines the type: skip, north-south corridor, east-west corridor, or room. The length of a corridor is the remainder modulo X or Y, as appropriate. The dimensions of the room are $width = R \bmod 4 + 2$ and $height = (R \ div \ 4) \bmod 4 + 2$ to yield dimensions for a rectangular room with both side lengths uniformly distributed in the range $2 \leq height, \ width \leq 5$. The upper-left corner (A, B) of the room or corridor are computed from the second integer N by computing $A = N \bmod X$ and

$B = (N \ div \ X) \ mod \ Y$ where the division is integer division. The "skip" type for objects makes it easy for mutation to *remove* features from a maze. Figure 2.5 gives an example of unpacking a gene of the sort used by this representation.

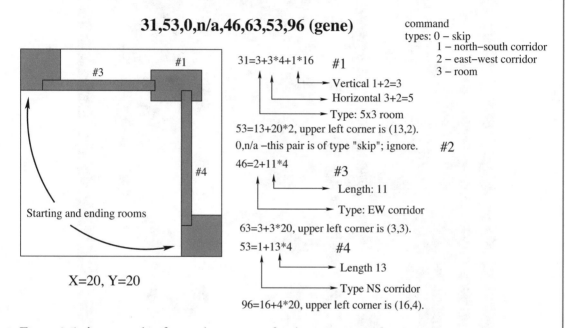

31,53,0,n/a,46,63,53,96 (gene)

command
types: 0 – skip
1 – north–south corridor
2 – east–west corridor
3 – room

31=3+3*4+1*16   #1

Vertical 1+2=3
Horizontal 3+2=5
Type: 5x3 room

53=13+20*2, upper left corner is (13,2).

0,n/a –this pair is of type "skip"; ignore.   #2

46=2+11*4   #3

Length: 11

Type: EW corridor

63=3+3*20, upper left corner is (3,3).

53=1+13*4   #4

Length 13

Type NS corridor

96=16+4*20, upper left corner is (16,4).

#3 #1 #4 #2

Starting and ending rooms

X=20, Y=20

Figure 2.5: An example of unpacking a gene for the negative, indirect representation with eight numbers (31, 53, 0, n/a, 46, 63, 53, 96) specifying four objects, one of which is ignored. The number n/a is not specified because it is part of a pair of type "skip." The removed rooms and corridors are shown as discrete, overlapping objects in the figure to illustrate the order of their application. In practice, the removed material would be represented as unobstructed grid squares. The 4 × 4 rooms placed at the entrance and exit are shown in this figure.

Two variation operators are used: uniform crossover with probability p_c and uniform mutation with probability p_m. The uniform crossover operates on two chromosomes, exchanging integers at each location with independent probability p_c. Uniform mutation operates by replacing the integer at each location with a new integer chosen at random in the range $0 \le n \le 9999$ with probability p_m. Actual rates for these operators, the number of pairs of integers in a chromosome, and the maximum number of filled squares Lim are given in the description of experiments in Section 2.6. An example of a maze evolved with the negative, indirect representation is shown in Figure 2.6.

The problem of most random mazes being untraversable did not arise with mazes encoded with the negative generative representation to anything like the degree it did with the direct and positive generative representations. For this reason no form of sparse initialization was used

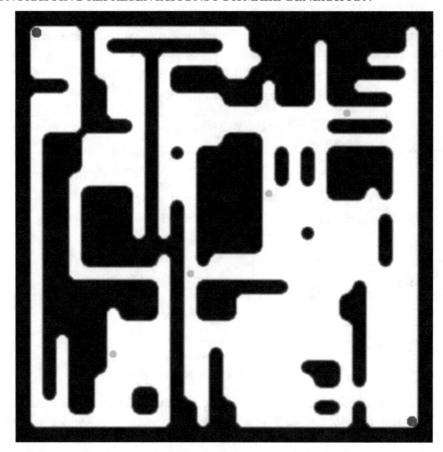

Figure 2.6: An example of a negative, indirectly encoded evolved maze. The larger red circles mark the entrance and exit while the smaller green ones mark checkpoints.

with the negative representation. This is another example of a change of representation causing change in the behavior of evolutionary search.

This also speaks to the inclusion of the skip type. In order to get a connected graph at all, this representation is allowed a generous collection of rooms. This means that the initial, random chromosomes are *too* well connected. Mutation to a skip type can perform evolutionary trial removal of rooms. In a way this strategy is the opposite of sparse initialization; it is fat initialization with the potential to slim down the connectivity of the level later.

2.5   FITNESS FUNCTION DESIGN

This section specifies elements from which fitness functions for maze-like levels can be built and specifies fitness functions used in the experiments in this study. The key to the system presented for constructing fitness functions are the presence of *checkpoints*. A checkpoint is nothing more than a position on the grid used to build the maze. Noting which checkpoints are on the shortest path from entrance to exit permits substantial control over what features of a maze are rewarded by a fitness function. The dynamic programming algorithm notes, for each grid square that is currently the tip of a shortest path, which checkpoint(s) have been encountered along any shortest path from the entrance to that square. When a new shortest path (tying the length of an existing path) is discovered leading to a grid square, any checkpoints discovered along that path are also added to the checkpoints recorded for that square. This recording of checkpoints within the dynamic programming algorithm permits membership and the various sorts of reconvergence defined below to be computed.

2.5.1   DEFINITIONS

The mazes in this study are assumed to have a single entrance square and a single exit square, typically in the upper left and lower right corner, respectively. This constraint is not difficult to relax in practice but yields a far cleaner mathematical environment for exploring fitness functions. Mazes are defined on a rectangular grid of squares in this study.

Definition 1 *The entrance (or start) square will be denoted as s, and the exit (or end) square will be denoted as e.*

Definition 2 *The set of* checkpoints *of a grid is a predetermined subset of the squares of the grid, this set is denoted by C.*

Definition 3 *The length of a minimal length path between any square x and the entrance square is denoted as $|x|$. This is also known as the* distance *between x and s. If this distance does not exist (because there is no unobstructed path from s to x) we set $|x| = -1$.*

Definition 4 *A checkpoint is a* member *of a square x if it is on a minimal length path from the entrance square to the square x.*

Definition 5 *The set of members of a square x is denoted by $\mathcal{M}_x$. The definition of the set of members of x yields a compact method of specifying which checkpoints share a particular relation with a square.*

Definition 6 *A square x is a* reconvergence *of checkpoints c_i and c_j if both c_i and c_j are members of x. Notice that a reconvergence is a common point along the shortest paths through multiple checkpoints. An example appears in Figure 2.7.*

Definition 7 *The* primary reconvergence *of checkpoints c_i and c_j is the smallest path length from the entrance square to any square which is a reconvergence of checkpoints c_i and c_j if a reconvergence exists, otherwise it is 0. The primary reconvergence value is denoted as $prc(c_i, c_j)$. We define this quantity to measure when paths from the entrance through pairs of checkpoints first happen to converge.*

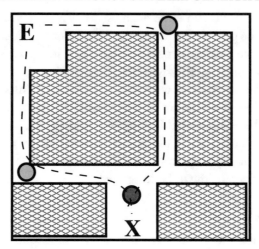

Figure 2.7: If the two green disks are checkpoints, then the square containing the red disk is an example of a place where paths from entrance **E** through the checkpoints to exit **X** re-converge.

Definition 8 *Any square y which is a reconvergence of checkpoints c_i and c_j whose distance to the entrance square is equal to the primary reconvergence of c_i and c_j is said to be a witness to the primary reconvergence of c_i and c_j.*

Definition 9 *An isolated primary reconvergence of checkpoints c_i and c_j is a witness to the primary reconvergence of c_i and c_j which has no checkpoints other than c_i and c_j as members. The function $iprc(c_i, c_j) = 0$ if no isolated primary reconvergence of c_i and c_j exists, and $iprc(c_i, c_j) = prc(c_i, c_j)$ otherwise.*

Definition 10 *A square x is a cul-de-sac if both $|x| \geq 0$ and for all adjacent squares y, $|x| \geq |y|$.*

Definition 11 *The set of all culs–de–sac in a maze is denoted by $\mathcal{Z}$.*

2.5.2   FITNESS FUNCTIONS

Assume the fitness function to be maximized is FF. We declare that $FF = 0$ if the entrance or exit squares are obstructed or if the exit square e has fewer than k members. The parameter k, the *mandatory exit checkpoint membership level*, gives an interesting form of control over the types of maze that evolve, as we will see in Sections 2.6 and 2.7. If e has k or more members, then $FF = F_i$ for one of the F_i given below.

- Exit Path Length Fitness

$$F_1 = |e|$$

The earlier study [7] found that this fitness function encourages mazes consisting of a single, long, winding path.

- Primary Reconvergence Sum Fitness.

$$F_2 = \sum_{x,y \in C} prc(x, y)$$

This fitness function strongly encourages accessibility of all checkpoints with the nature of this accessibility controlled by k. If a checkpoint is not required to be on a shortest path to the exit then higher fitness results from placing it at the end of a cul-de-sac.

- Isolated Primary Reconvergence Sum Fitness

$$F_3 = \sum_{x,y \in C} iprc(x, y)$$

Isolated reconvergence is harder to achieve than simple reconvergence. It strongly encourages that paths branch, run over checkpoints, and then meet up again later.

- Cul-de-sac Count Fitness

$$F_4 = |\mathcal{Z}|$$

- Cul-de-sac Length Fitness

$$F_5 = \sum_{x \in \mathcal{Z}} |x|$$

2.6   DESIGN OF EXPERIMENTS

A very similar evolutionary algorithm was used for all experiments. The algorithm is steady state [67] using a size seven single tournament selection. The tournament size sets the selection pressure in the algorithm and seven is an intermediate value that creates neither particularly strong or weak selection pressure. If you are implementing ideas from this text, you may want to experiment with varying the size of the tournaments. The steady-state model of evolution proceeds by mating events that generate pairs of new structures that are reinserted into the population before the next mating event. Size seven single tournament selection chooses seven members of the population without replacement. The two best are copied over the two worst. The binary variation operator (crossover) is applied to the two copies, then the unary variation operator (mutation) applied to each of the copies. The algorithm is run for 500,000 mating events, saving summary fitness statistics for the population every 2,000 mating events. Such a block of 2,000 mating events is called a *generation*. Each experiment consisted of 30 replicates of the evolutionary algorithm performed with distinct random number seeds. The crossover and mutation rates are set to $p_c = 0.05$ and $p_m = 0.01$.

Both of the direct representations and the positive generative representations use a population size of 120, while the negative generative representation uses a population size of 1,000.

The direct and positive representations used sparse initialization, while the negative representation did not: its larger population size makes it very likely that mazes with positive fitness are present in the initial populations for the negative representation.

For all representations, both sparse initialization and large populations were tested for their ability to yield an initial population with a majority of feasible mazes. Which of these two options was found to be better was different for different representations. Note that since the negative representation removes material from an initially full grid, sparse initialization would require a more complex structure rather than a simpler one as is the case in the other three representations.

Since our purpose is to make a qualitative comparison of the types of maze evolved with the three representations *rather* than to compare their fitness, this rather substantial difference in initialization technique and population size is not a problem: each algorithm was tuned to produce an initial population with solvable mazes, *not* to compare the representation's ability to solve the same optimization task.

2.6.1   INITIAL EXPERIMENTS

A set of nine initial experiments were performed using fitness functions F_1–F_3 with each of the three non-chromatic representations. These all used a 30×30 grid for the maze with checkpoints set $C = \{(6, 24), (12, 18), (18, 12), (24, 6)\}$. The entrance is square $(0,0)$, while the exit is square $(29,29)$. The mandatory exit checkpoint membership level was set to $k = 3$.

2.6.2   EXPERIMENTS WITH CULS-DE-SAC

A pair of experiments were conducted, using the binary direct representation, to test fitness functions F_4 and F_5, which maximize the number or distance from the entrance of culs-de-sac. These experiments use the binary direct representation with the same settings and checkpoints as the other experiments that use the binary direct representation. As with the other experiments some requirements are placed on the maze: there must be a path from the entrance to the exit and all checkpoints must be in unobstructed squares.

2.6.3   CHANGING THE BOARD SIZE

Three experiments, using identical algorithm settings to the initial experiments for all three non-chromatic representations using fitness function F_3, were performed on a 50×20 grid with checkpoints $C = \{(5, 15), (15, 15), (5, 35), (15, 35)\}$. These experiments are supposed to demonstrate the ability of the algorithm to work for non-square boards and boards of other sizes.

2.6.4   EXPERIMENTS WITH THE CHROMATIC REPRESENTATION

Four experiments were conducted, consisting of 30 replicates each, with the chromatic representation. These experiments differed from one another in using fitness functions F_1, F_2, F_3, and F_4. These experiments use a 30×30 grid and otherwise use the same algorithm settings as the initial nine experiments, except that the required number of checkpoints that must be members of the exit square were reduced to $k = 1$. These experiments are intended to see if the various fitness functions used in this study work well with a representation that specifies barriers implicitly.

2.6.5 VERIFICATION OF SPARSE INITIALIZATION AND CROSSOVER

The sparse initialization and the choice of a low-rate uniform crossover were both based on preliminary experimentation with a development version of the software created for this study. Four additional experiments were performed, based on the initial experiment with the binary direct representation using F_1. Two of these experiments replace uniform crossover with one-point and two-point crossover. The other two increased the *fill* parameter from 0.05 to 0.1 or 0.2. Recall that this parameter governs the probability that a square in the binary direct representation be obstructed.

2.7 RESULTS AND DISCUSSION FOR MAZE GENERATION

Figure 2.8 shows a maze from each of the nine original experiments. Up to the limitations imposed by the representation, mazes evolved with fitness function F_1 which rewards only the length of the path from the entrance to the exit produced long winding paths with little or no branching and only a few side corridors. This is true of all 30 replicates for each of the 3 experiments using F_1. The effect of the k parameter, when this fitness function is used, is to force the checkpoints to be on this path. This means a designer can use checkpoints to shape the path to some degree.

For the mazes generated using F_2, the requirement that three checkpoints be members of the exit comes into play. If none of the checkpoints are required to be on a shortest path to the exit, then it is very likely to have one checkpoint on such a path and place the others at the end of long culs-de-sac. This maximizes the reconvergence numbers for pairs of checkpoints very well. When we require, by setting $k > 1$, that several checkpoints be on a path from the entrance to the exit, then the algorithm places the required number of checkpoints on paths from entrance to the exit and places remaining checkpoints at the end of culs-de-sac. In all three of the mazes shown in Figure 2.8 the lower left checkpoint is at the end of a long cul-de-sac. The mazes evolved with the negative and binary direct representation place the checkpoints in a branching structure, while the positive representation places all three on a long, winding path.

All three of the mazes evolved with F_3 place all four checkpoints on distinct paths to the exit. The pattern of branching varies somewhat and both generative representations produce more additional large branches. This fitness function was the most likely, during algorithm development, to produce entire populations that are all of zero fitness. It also is not as strongly

Figure 2.8: Shown is a maze from each of the nine initial experiments. The mazes are organized so that all mazes in a column use the same representation and all mazes in a row use the same fitness function. The large, red circles mark the entrance and exit while the small, green dots mark the checkpoints. These mazes are 30×30 and the entrance, at $(0,0)$ is in the upper left corner while the exit, and at $(29,29)$ is in the lower right.

impacted by the value of k as the need to produce unique reconvergences to obtain fitness forces checkpoints to be on direct paths from the entrance to the exit. A checkpoint in a cul-de-sac can give rise to *at most* one unique reconvergence, while one along a path from entrance to exit can have a unique reconvergence with every other checkpoint.

The different population sizes chosen for experiments with different representations were the result of initial experimentation with the goal of finding a population size that permitted the algorithm to generate acceptable results. In [27] it is shown that, for some problems, very small population sizes are superior while others function better with large populations. The relationship between final quality and initial population size is complex, obscure, and representation dependent. Since the goal of this study is to show what different representations and fitness functions can do, we did not perform an extensive parameter study to locate good initial population sizes and rather used a quick *ad hoc* set of trials.

The binary direct representation produces almost "intestinal" patterns that are more reminiscent of a natural cave than the results produced by the two indirect representations. The positive representation generated the sparsest maps as well as those that looked like intentional or planned structures. The negative representation produced mazes with a distinct character from the other two, but hard to describe stylistically. Examination of the full range of evolved mazes shows that the negative representation was the best at placing checkpoints on distinct paths from the entrance to the exit; the direct representation was second best by this criterion; and the positive representation came in a distant third in this regard.

In the initial stages of the research reported in [17], a version of the direct representation was tried and failed horribly: no path from entrance to exit was found in almost all mazes in the initial population. In this study the use of sparse initialization (in both the direct and positive representations) solved this problem. In this case *sparse initialization* means initialization to a state of the maze where relatively few squares are obstructed. The decision to use low-rate uniform crossover in this study was made before sparse initialization was implemented, as a means of enhancing the heritability of feasible mazes. When this choice was later revisited for the binary direct representation in a final experiment, reported in Section 2.7.3, we found that the performance of standard crossover is similar to but significantly superior to the uniform crossover used in most of the experiments.

The use of sparse initialization is interesting to consider from the perspective of a fitness landscape as well. Sparse initialization was used to create a situation in which there are a very large number of different paths, on average, from entrance to exit and from either entrance or exit to each of the checkpoints. This means that minimal path length and all the reconvergence scores are fairly small. In essence, we intentionally create solutions that are very likely to be (i) low fitness and (ii) feasible. This means that we are relying on the variation operators, guided by selection, to climb the hills of the adaptive landscape. Since high fitness mazes are typically very close, when distance is measured in mutations, to infeasible mazes this is probably an effective strategy. Examine the maze in Figure 2.8 for the direct representation and F_1 (which simply maximizes the path length from entrance to exit). Inspection shows a *majority* of the unobstructed locations will block the path from entrance to exit if they are filled in.

If we look at the mazes generated with F_3, which encourages large, isolated primary reconvergences, we see that no single loci in the gene will *block* the path from entrance to exit. Inspection shows that several loci can zero out as many as three of the isolated reconvergences if they are mutated. This means that for F_3, even though mutations to zero fitness are uncommon, mutations to much lower fitness are common in high-fitness, non-sparse structures. We thus see that sparse initialization also places chromosomes in a part of the fitness landscape that is far less rugged than the regions containing high-fitness structures.

Examining Figures 2.9–2.11 we see that all nine sets of experiments experienced substantial optimization of the fitness function: verifying this is one of the reasons for using a mean-over-replicates fitness plot of this type. Several of these plots, particularly the one using F_3 and the negative representation, suggest that there is still an upward trend. This is not a great concern; the algorithm can easily be run for a longer time. Of greater importance is the fact that it is not clear that the global optimum of any of these fitness functions is the most desirable maze or level design. These algorithms are supposed to provide a broad selection of mazes and the fitness functions are rough heuristics, not clear statements of entirely desirable objectives. Figure 2.12 shows the behavior of population average fitness in a single run. Notice that the major improvements are via a type of innovative leap (recall a "generation" consists of 2000 mating events of steady state evolution) rather than steady progress. In addition, the increase in the width of the confidence intervals on mean fitness as mean fitness improves supports the earlier assertion that high fitness parts of the fitness landscape are mutationally near to cliffs. The innovation represents climbing such a cliff while the high variation in mean fitness represents individuals that have fallen off such cliffs.

2.7.1   EXPERIMENTS WITH CULS-DE-SAC

Figure 2.13 shows an example maze evolved with each of F_4 and F_5. The F_4 maze optimizes the number of culs-de-sac in the maze. This, together with the requirement that the checkpoints appear in the area accessible from the entrance, yields a maze with a large number of loops and side passages relative to those produced with the other fitness functions. The function F_5 sums the size (distance from the entrance) of all the culs-de-sac. It produces mazes similar to those produced by F_1, maximizing the length of the path from entrance to exit, but with more and longer side passages.

The F_4 mazes are, upon cursory inspection, the most complex and diverse of the sets of mazes evolved with the various fitness functions presented. This suggests that investigation of fitness functions that count culs-de-sac is worth more attention.

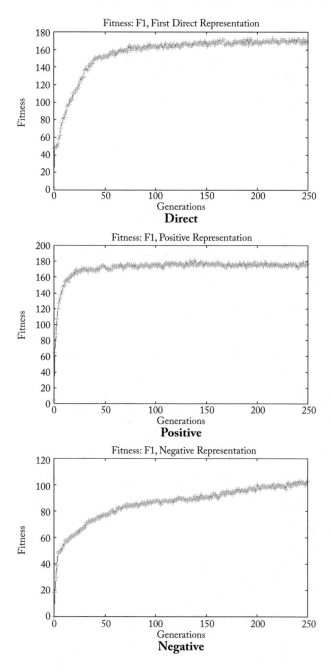

Figure 2.9: The plots give mean fitness across evolution, for all 30 replicates, for each of the 3 representations on the first fitness function.

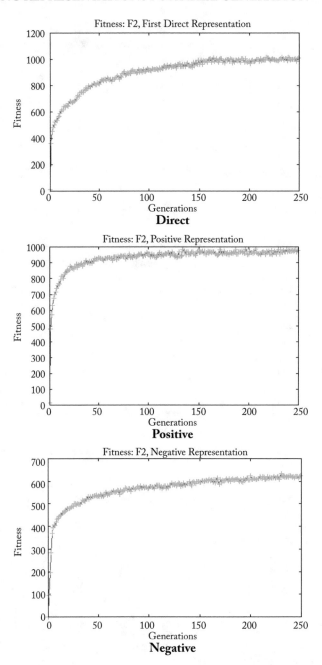

Figure 2.10: The plots give mean fitness across evolution, for all 30 replicates, for each of the 3 representations on the second fitness function.

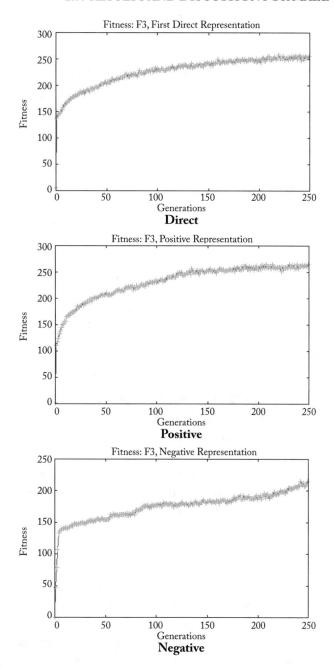

Figure 2.11: The plots give mean fitness across evolution, for all 30 replicates, for each of the 3 representations on the third fitness function.

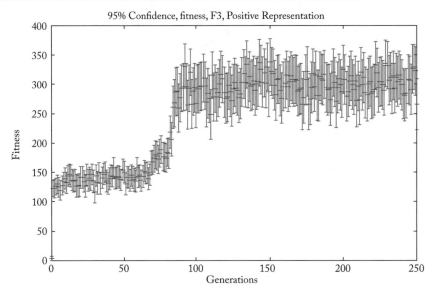

Figure 2.12: Shown is the mean fitness over the course of evolution, in the first evolutionary replicate, using fitness function $F3$ with the positive representation. Shown with confidence intervals representing a 95% confidence interval.

2.7.2   EXPERIMENTS WITH DIFFERENT BOARD SIZES

These experiments represented a check on the ability of the algorithm, for all three non-chromatic representations, to work on a different board size and with different arrangements of checkpoints. Examples of mazes for each representation are shown in Figure 2.14. A visual inspection of the 30 mazes produced in each experiment (data not shown) shows that the algorithm is slightly more likely to place checkpoints at the end of long culs-de-sac when the board has a 5:2 aspect ratio, but otherwise the results were similar to those obtained for the 30×30 grid.

Chromatic Mazes

Figure 2.15 shows an example of a chromatic maze and its key from each of the four experiments. The initialization to green and yellow colors is most visible in the mazes evolved with F_4. The other fitness functions all yield a fairly even distribution of colors.

The resulting mazes have similar characters to those evolved with the other three representations. The maze evolved with F_1 is very long but can be solved by mere persistence; it has few side branches and no real choices other than forward and back. The mazes created with F_2 and F_3 place checkpoints not required to be on a shortest path to the exit at the end of long culs-de-sac. The mazes created with F_3 have more branches than those created with F_2.

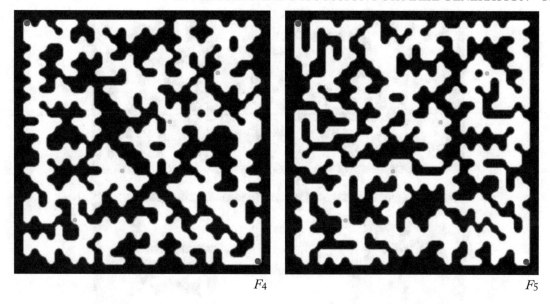

F_4 F_5

Figure 2.13: Example mazes for fitness functions F_4 and F_5 using the binary direct representation. The algorithm that produced these mazes also requires that all four checkpoints appear in the part of the maze accessible from the entrance. The large red circles mark the entrance and exit, while the small green ones denote checkpoints.

The mazes created with F_4, which rewards having culs-de-sac, creates a maze with the largest number of branches, closed loops, and reconverging paths. This is also consistent with the behavior of F_4 in the experiments run with the other representations.

Contrasting the chromatic mazes, rendered directly, with their keys shows that the implicit character of the barriers between squares makes this type of maze much harder to solve than one where the barriers are clearly visible. The chromatic representation is one of a large collection of possible representations for implicitly specified mazes. The experiments in this study show that the space of fitness functions defined here can be applied to implicitly specified mazes.

2.7.3 SPARSE INITIALIZATION AND CHOICE OF CROSSOVER OPERATOR

Table 2.1 and Figure 2.16 show the results of varying the crossover operator for the binary direct representation using F_1. The initial choice of the low rate uniform crossover was made while experimenting with the problem of initial populations with zero fitness. One-point crossover exhibits slightly higher performance than uniform crossover, but the difference is not significant. Two-point crossover outperformed both uniform and one-point crossover significantly. The margin of improvement is small, $14.4/362.5 \cong 4.0\%$. Figure 2.16 shows that none of the

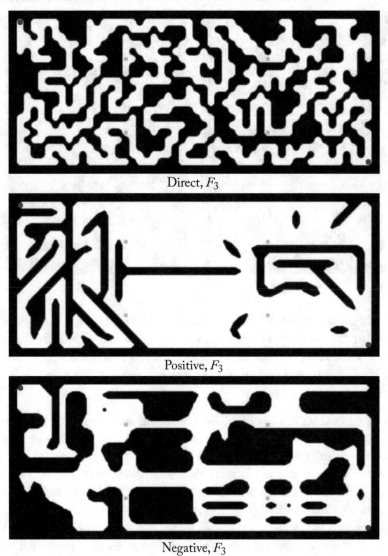

Direct, F_3

Positive, F_3

Negative, F_3

Figure 2.14: Examples of 50×20 mazes for all three non-chromatic representations. Large red circles mark the entrance and exit while small green ones mark the checkpoints.

different crossover operators were particularly faster at reaching the final fitness level. This suggests that the choice of crossover operator is not important, but that future work with these representations and similar representations should use two-point crossover.

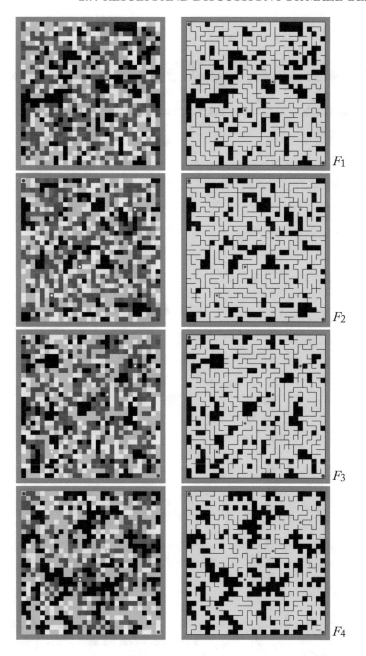

Figure 2.15: Chromatic mazes and keys. Black squares are inaccessible from the entrance. Smaller red circles denote checkpoints, larger gray ones the entrance and exit of the maze. The fitness function used to evolve each maze is given to the lower right of its key rendering.

Table 2.1: Shown are 95% confidence intervals for experiments with the binary direct representation using F_1 with different crossover operators

Crossover Operator	Fitness	95% Confidence Interval
Uniform	362.5 ± 4.0	(358.5,366.5)
One-Point	365.1 ± 2.5	(362.6,367.6)
Two-Point	376.9 ± 2.6	(374.3,379.5)

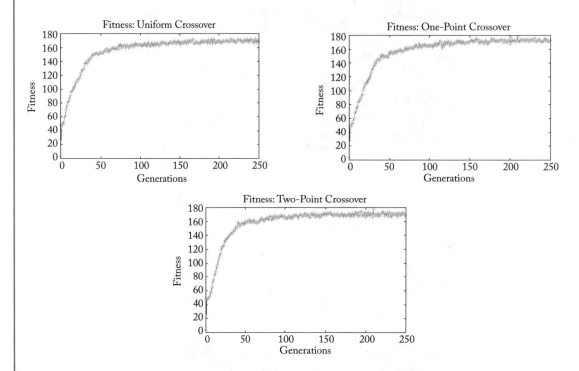

Figure 2.16: Mean fitness over 30 replicates for uniform, one-point, and two-point crossover, as a function of number of generations. A generation consists of 2,000 mating events.

The results of varying the *fill* parameter were more dramatic. When the fill parameter was set to 0.05, the standard setting, all 30 replicates produced a maze with positive fitness. When *fill* = 0.1, 17 of 30 replicates produced a feasible maze. When *fill* = 0.2, none of the 30 replicates produced a feasible maze. This result strongly supports the use of sparse initialization for the binary direct representation. It also tends to support its use based on early experimentation with the chromatic representation and the indirect, positive representation.

It might be natural to ask why not just set *fill* = 0, starting with an empty grid and letting the variation operators do all the work. One of the factors that permits evolutionary algorithms to function is the random initialization. This initialization and the subsequent shake-out of the population acts to select the basin of attraction of the dynamical system represented by the evolutionary algorithm in which a given replicate arrives at its final solution. Even a modest amount of randomization, such as that represented by a setting of *fill* = 0.05 is likely to substantially increase the diversity of mazes located by the EA.

2.7.4   ALGORITHM SPEED

Running a set of 30 replicates for 500,000 mating events requires about as much time as a coffee break (8–20 m). The actual timing is somewhat variable, depending on the fitness function used. Oddly, this is *not* because the different fitness functions require different amounts of time to compute. The current version of the algorithm computes all the elements required by any of the fitness functions and then performs a trivial computation to return the exact fitness function being tested. This was done to permit rapid exploration of the space of fitness functions and to leave a hook for future research on multicriteria optimization based versions of this software. The bottleneck in the fitness function is running the dynamic programming algorithm. The mazes of the sort that get a high score from F_4 require that far more squares be visited (with repetition for multiple paths from the entrance to a square) than the mazes that evolve under the influence of F_1. This phenomena is most apparent when sparse representations are used: dynamic programming runs *very* slowly on an almost empty maze. If the researcher watches the trace of a run, the rate of data reports jumps upward as the fitness increases.

The current version of the algorithm is more than adequate for producing huge libraries of mazes with similar properties but different details. Optimizing to compute only the factors needed in a given fitness function will yield a small increase in speed, but the real point to look for speed improvement is the dynamic programming algorithm. The algorithms used in this study were designed to make explorations of the space of fitness functions easy rather than being optimized for speed. Substantial research exists on optimizing dynamic programming algorithms, and the algorithm used here can be parallelized trivially by running one copy of the evolutionary algorithm on each processor. In addition, usable mazes arise long before 500,000 mating events. Optimizing the effort/return trade-off of evolution is another area where additional speed may be obtained.

2.7.5   FITNESS LANDSCAPES AND SPARSE INITIALIZATION

Both direct representations and the positive indirect representation required a form of sparse initialization to achieve adequate performance. The effect of this type of initialization is to create mazes with very few obstructions. A maze with few obstructions is likely, for all five fitness functions, to have low positive fitness. In particular, zero fitness because a lack of any path from the entrance to exit is avoided by sparse initialization. This, in turn, makes the variation operators bear the brunt of the burden of generating high fitness mazes. The experiments show that they do this effectively. The experiments in which the *fill* parameter was varied show that sparse initialization is necessary.

Examine Figure 2.12. This figure shows that, in one run, as the mean fitness of the population increases so does the variance of the fitness. Examining the evolved mazes shown, it is obvious that there are many squares which, if obstructed, will cause a catastrophic decrease in fitness. These two pieces of evidence demonstrate that high fitness mazes are likely to have mutants with much lower fitness. Colloquially, the heights of the fitness landscape have many cliffs. The sudden increase in fitness shown in Figure 2.12 is probably driven by a sudden jump *up* one of these cliffs. This "high fitness implies instability" may be an example of a type of self-organized criticality. It also provides a reason, in addition to *ad hoc* observation of unfit initial populations and the experiments varying *fill* with the binary direct representation, why sparse initialization may be a good idea.

2.7.6   DISCUSSION FOR MAZE GENERATION

The primary contributions in this chapter are threefold. First, it is demonstrated that four different representations can be used, with the same fitness functions, to generate maze-like levels for use in games. The representations generate mazes with very different appearances and characters.

Second, this study defines several elements, computable with a simple dynamic programming algorithm, that can be used to build a large number of different fitness functions. Five such fitness functions are tested in this study, and many others can be built from primary reconvergence, isolated primary reconvergence, path length from entrance to exit, number of culs-de-sac, path lengths of culs-de-sac, number of checkpoints that are accessible, and number of checkpoints on shortest paths from the entrance to the exit of the maze. The study also demonstrates that the mazes generated with different fitness functions are substantially different from one another in terms of branching factors, loops within the maze, and the placement and length of culs-de-sac.

Third, the study demonstrates that the fitness functions yield comparable results on all four representations tested, including the chromatic representation which specifies a maze implicitly rather than explicitly.

The key rendering of the chromatic representation is a tool for apprehending the connectivity and solubility of the implicitly represented maze in an explicit form. The use of multiple renderings, some of which are available only to a designer, is a potentially rich area. Processing the data from the dynamic programming algorithm in various ways could yield many views useful to a designer.

Sparse initialization, used in all the representations except the negative indirect representation, is a simple but potentially valuable technique for use in the implementation of the representations presented in this study and, potentially, others. Its value was directly verified for one representation and fitness function and is suggested by results during algorithm development for the other two representations where it is used.

While exact quantization of the optimized speed of procedural content generation via the methods prototyped in this study remains to be done, the study demonstrates that procedural generation of a variety of different types of maze-like maps can be performed rapidly. For off-line generation of libraries of content, the techniques in this study are already beyond the speed required for application. Real-time content generation is certainly within the realm of the possible once the algorithms have been optimized.

The most obvious next step for this research is to build a GUI tool for designing maze-like game levels. In this context the algorithm would act as a designer's assistant. A dialog box for piecing together a fitness function out of the elements described here would permit a designer a great deal of freedom to control the types of maze that evolve. The number of fitness functions *not* explored in this study is immense. As researchers found new elements these could be added to the list of options. A palette of representations, including but not limited to those presented here, would increase the flexibility of the tool.

A clear direction in which to extend this research it to find other fitness function elements that could be used to increase the reach of the techniques. Elements that are easy to compute from the variables present in the dynamic programming algorithm used in this study would be most desirable. One such element would be the number of inaccessible squares in a maze, either in absolute terms, or that are nominally unobstructed but not accessible from the entrance of the maze.

The use of checkpoints, while key to defining many of the fitness function elements used, was not extensively explored here. The mazes presented during demonstration of the algorithms used four checkpoints and only two different arrangements of those checkpoints were tested. Varying the number and arrangement of checkpoints should yield a great deal of control over the type of mazes that arise. That work is left for the future; it is expected that there will be some unexpected consequences of moving checkpoints and varying the number of checkpoints used. Likewise, changing the k parameter, denoting the number of checkpoints that must be a member of the exit square, requires more exploration.

While several representation were tested in this study, many others are possible. Grammatical systems, particularly L-systems [47, 58], seem a natural candidate. We are aware of no research on generating mazes with L-systems. Our group has worked with L-systems for other purposes [9, 10, 26] and our intuition is that the sort of constraint handling performed with dynamic programming would be difficult to achieve with L-systems. Nevertheless, L-systems remain a technology of interest, in particular because a successful L-system representation would generate levels with remarkable speed. We note that Parish and Muller [56] have used L-systems to design road networks, a similar task to maze design.

Another factor that would work well with the dynamic programming based fitness functions presented in this study is the incorporation of measures of area filled. At its simplest, this could consist of noting how many squares of a maze grid were filled in. More sophisticated measures could test for the presence of open areas of at least a given shape or of a particular size.

2.7.7 BREAKING OUT OF TWO DIMENSIONS

This book is an initial survey and so neglects some topics. The representations in this chapter can be extended to create three-dimensional mazes or, if we wanted to add teleporters, mazes with even more dimensions. This topic is an interesting and important one that we regretfully lack the time to delve into.

CHAPTER 3

Dual Mazes

This section explores the creation of *dual mazes* which have multiple barrier types and, using them, create multiple mazes occupying the same physical space. A maze using barriers made of fire, water, and stone is shown in Figure 3.1 with a key map showing the connectivity of the stone and fire and water and fire mazes appears in Figure 3.2.

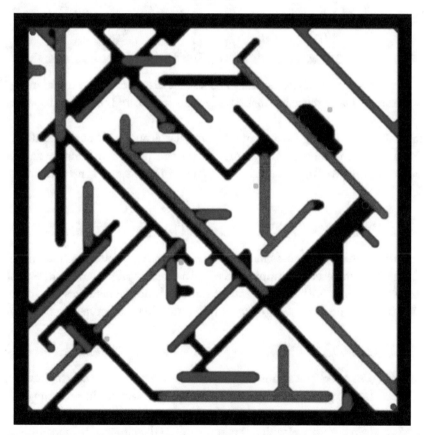

Figure 3.1: An evolved maze with fire (red), water (blue), and stone (black) obstacles. The maze defined by the water and stone was evolved to maximize the number of culs-de-sac, while the maze defined by fire and stone was evolved to maximize path length from entrance to exit. Green dots indicate checkpoints while the small red circles mark the entrance and exit.

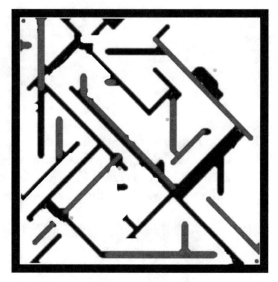

Figure 3.2: This is a key map, showing the to co-extant mazes depicted together in Figure 3.1. Notice that the stone-water maze is substantially more connected than the stone-fire maze.

A map defined by having squares in a rectangular array (grid) at different elevations appears in Figure 3.3. In this latter map some agents can jump a maximum distance of 1 m, while others can jump two. The generation of maps of these types is a more complex version of SBPCG.

3.1   REPRESENTATIONS FOR DUAL MAZE GENERATION

Two representations are used in this chapter: a generative representation that specifies barriers to be added within an enclosed but empty space and a direct representation that specifies terrain heights at each square of a grid. The generative representation generalizes the indirect positive representation from the previous section by adding multiple barrier types to the representation.

3.2   DETAILS OF THE GENERATIVE REPRESENTATION

As before, a collection of specifications for barriers is stored as a linear array of integers in the range $0 \ldots 9999$. The integers are used in pairs to specify barriers in an empty $X \times Y$ arena with walls at its edges. The first integer is bit-sliced into a one-bit number, a three-bit number, a trinary number, and a remainder R which is taken modulo the larger of the two arena dimensions to yield the length of the barrier. The one-bit number determines if a barrier is penetrating or not. A penetrating barrier runs from its starting point to its length or an edge of the arena. A non-penetrating barrier stops when it encounters any other barrier. The three-bit number is interpreted as a direction for the barrier to run from its starting point W, NW, N, NE, E, SE, S,

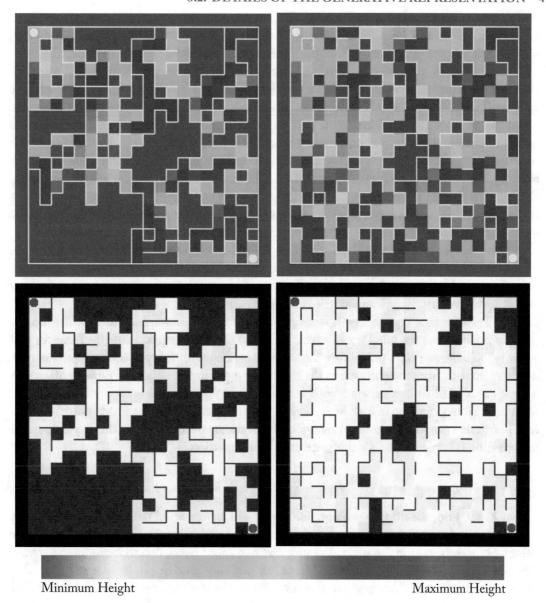

Minimum Height   Maximum Height

Figure 3.3: A map defined by variable height terrain. The left upper panel shows squares accessible to an agent that can only jump one meter while the right upper panel shows squares accessible to an agent that can jump two meters. White lines indicate height differences too great for a jump between adjacent squares and these are keyed more visibly in the lower panels. Agents enter the maze at the yellow circle in the upper left and their goal is the yellow circle in the lower right. Height is given by HSV color. The lower panels show key maps for the maps in the upper panels.

or SW. The trinary number specifies stone, fire, or water for the type of the barrier. The second integer i is split as $x = i \bmod X$ and $y = ((i \ div \ X) \bmod Y)$ to obtain the starting point (x, y) of the barrier on the grid. The operator *div* represents integer division without remainder.

An additional parameter, an integer, is used in the generative representation. A given instance of the representation is permitted to lay down at most a fixed number, *Lim*, of obstructed squares. Once this number of filled squares is reached, the remainder of the barrier being laid down and all other barriers following it are skipped. Two variation operators are used: uniform crossover with probability p_c and uniform mutation with probability p_m. Uniform crossover operates on two chromosomes, exchanging integers at each location with independent probability p_c. Uniform mutation operates by replacing the integer at each location with a new integer chosen at random in the range $0 \ldots 9999$ with probability p_m. Actual rates for these operators, the number of pairs of integers in a chromosome, the *representation length*, and the maximum number of filled squares, *Lim*, are given in the description of experiments in Section 3.3.

An example of a maze evolved with the generative representation using multiple barrier types is shown in Figure 3.1. The evolutionary algorithm evaluates two mazes derived from such a specification. We assume that one type of agent is fireproof (i.e., can traverse red squares) but unable to touch water (i.e., cannot traverse blue squares) so that the water and stone barriers define a maze for that type of agent. A second type of agent cannot pass through the barriers made of fire but does not mind water so that the stone and fire barriers define the maze for that agent type. There is a high probability that a maze created with this generative representation will have no path from the entrance to the exit of the maze unless either the number of barriers is kept quite small or barriers start with very short lengths. A maze with few barriers is not likely to be interesting no matter how cleverly they are placed; for this reason initial barriers (but not those created by mutation) have lengths in the range 1–3. This is the way that sparse initialization is implemented for this generative representation.

3.2.1   DETAILS OF THE DIRECT REPRESENTATION

For an $X \times Y$ grid, the direct representation consists of an array of XY real values generated by a Gaussian random variable with standard deviation 1.0 and mean 3.0. These represent terrain heights, in meters, within the maze. We presume two agent types, one that can jump between adjacent squares whose heights vary by 1 m or less and another that can jump up or down 2 m or less. The barriers in this maze are thus implicit in the distribution of terrain heights. The two agents, one-meter jumpers and two-meter jumpers, encounter different barriers, and so the specification of terrain heights defines two mazes with the height one maze nested inside the height two maze.

Two variation operators are used: uniform crossover with probability p_c and uniform mutation with probability p_m. The uniform crossover operates on two chromosomes, exchanging their bits at each location with independent probability p_c. Uniform mutation operates by adding a Gaussian random number with a specified standard deviation to the height at each location with probability p_m. Actual rates for these operators are given in the description of experiments in Section 3.3. An example of a maze evolved with this representation is shown in Figure 3.3.

This representation also requires a form of sparse initialization: the heights are initially set to 3 m plus a standard normal random variable (mean 0, standard deviation 1). A maze initialized in this fashion permits movement between many pairs of squares while making it very easy for mutation to create more barriers. This representation is very different from the generative representation in the following sense. The other representation specified which squares of a grid are obstructed by stone, fire, or water. The height-based direct representation does not directly obstruct any square within the grid: rather it imposes rules about which squares it is possible to move between. This means that it creates implicit walls between squares that define the maze rather than defining it with explicitly obstructed squares. Notice that there are far more locations to place an implicit barrier between squares of a grid than there are squares of a grid to obstruct. This suggests that the resulting mazes have the potential to be more complex.

3.2.2   FITNESS FUNCTION SPECIFICATION

The following is a list of the single-maze fitness functions used in this section. They use the definitions given in Chapter 2:

$$f_{path} = |e| \tag{3.1}$$

$$f_{cul} = |Z| \tag{3.2}$$

$$f_{prim} = \sum_{c_i \neq c_j \in C} prc(c_i, c_j) \tag{3.3}$$

$$f_{isoprim} = \sum_{c_i \neq c_j \in C} iprc(c_i, c_j) \tag{3.4}$$

Both representations used specify two distinct mazes. Each maze is evaluated with one of the fitness functions above, and then selection operates on the geometric mean, $\sqrt{f1 \cdot f2}$, of the two fitness values. This choice assumes that the two fitness values are roughly equally important and should have values as close as possible to equal. In contrast, the arithmetic mean, $\frac{1}{2}(f_1 + f_2)$, assumes that fitness from either function is equally valuable, in particular that a zero result from one function can be balanced by high values of the other. Given that we wish both fitness functions to have a moderately large positive value relative to their natural range, the arithmetic mean is not a good choice. It is clear that using multicriteria optimization would permit a greater diversity of mazes to be located, but a substantial diversity of mazes is located with this simple system. The application of multicriteria optimization to this problem is left for subsequent studies.

3.3   EXPERIMENTAL DESIGN

Two experiments were performed with the direct representation and four with the generative representation. Each experiment consisted of 30 replicates of the evolutionary algorithm. The population size was set to 120 for all 6 experiments. The evolutionary algorithm used is steady state with size seven single tournament selection used to perform selection. This model of evolution picks seven members of the population without replacement. The two best are copied over the two worst in the tournament, breaking ties uniformly at random, and then the variation operators are applied to the copies. In all six experiments the rate for uniform crossover was set to $p_c = 0.05$ and the rate for the uniform mutation operator was set to $p_m = 0.01$. The Gaussian mutation used with the direct representation used a standard deviation of $\sigma = 0.5$ and a mean of 0.

   Experiments with the direct representation used a 20×20 grid of heights represented as an array of 400 values, while the generative representation used a 40×40 grid with a representation length encoding 60 barriers. The set of checkpoints used is $\mathcal{C} = \{(8, 32), (16, 24), (24, 16), (32, 8)\}$ in the experiments with the generative representation and $\mathcal{C} = \{(4, 16), (8, 12), (12, 8), (16, 4)\}$ with the direct representation. It was required that all four checkpoints be situated within the part of the maze accessible from the entrance. A fitness of zero was awarded to a maze in which this was not so. A *mating event* is one instance of single tournament selection. Experiments with the direct representation continued for 5,000,000 mating events with summary fitness statistics saved every 2,000 mating events. Experiments with the generative representation continued for 2,500,000 mating events with summary fitness statistics saved every 1,000 mating events. The shorter evolutionary time for the generative representations was selected during testing of the system because fitness typically stopped increasing after 2,500,000 mating events.

   It remains to specify which fitness functions were used to drive selection. The fitness functions used in the six experiments are summarized in Table 3.1. In the direct representation it is clear that the squares in the maze that the agent with jump height one can reach are a subset of the maze the agent with jump height two can access. This means that the geometric average will cause Experiment 1 to generate a long path for the jump height one agent with some short-cuts for the height two agent. Experiment 2 looks at the results of using very different fitness functions to select the mazes seen by the two different types of agents.

   Experiments 3–6 used the generative representation. The parameter *Lim* that specifies the maximum number of squares that may be obstructed was set to 400 in these experiments, permitting up to 25% of the space in the 40×40 grid to be obstructed by barriers of any sort. Experiment 3 uses the same fitness function as Experiment 1 with a very different result, because the longest path in the two mazes are not forced to be subsets of one another and can, in fact, be very different. Experiments 4–6 use the culs-de-sac maximization fitness function for the fireproof agent and vary the fitness function for the agent that can cross water. The culs-de-sac fitness function was found to generate the most diverse collection of mazes in [13].

Table 3.1: Experiments listed by representation and types of fitness function used

Experiment Number	Representation Type	Fitness Function	
		Jump 1	Jump 2
1	Direct	f_{path}	f_{path}
2	Direct	f_{path}	f_{cul}
		Stone + Fire	Stone + Water
3	Generative	f_{path}	f_{path}
4	Generative	f_{path}	f_{cul}
5	Generative	f_{prim}	f_{cul}
6	Generative	$f_{isoprim}$	f_{cul}

The experiments in this section represent a proof-of-concept study for using the geometric mean of two different fitness functions for the embedded mazes within a dual maze. A complete exploration of the available representations (those already developed in this study and its predecessors) and all pairs of fitness functions would require $5 \cdot \binom{5}{2} = 50$ experiments. Those presented here were chosen, based on earlier work, as being the most likely to yield interesting (or in the case of Experiment 3, boring but informative) results.

3.4   RESULTS AND DISCUSSION FOR DUAL MAZES

Figure 3.4 shows the value of the two fitness functions used in one of the 30 replicates of Experiment 2. This plot is typical of similar plots in all the experiments; it serves to certify that evolution is functioning nominally as a search algorithm in the experiments.

Figure 3.5 shows example results from Experiments 1 and 2. Height is shown using an HSV color scale with white barriers indicating drops or rises that are too large for an agent to jump between adjacent squares. The mazes on the left show only those squares accessible to the agent with an ability to jump up or down 1 m; those on the right are similar but for the agent that can jump up or down 2 m. Grey squares are inaccessible to the agent. Sparse initialization to a height near the middle of the distribution of heights is responsible for the over-representation of colors near the middle of the height distribution.

In both Experiments 1 and 2 the fitness function used for the maze accessible to the height one agent was f_{path}, maximizing the path length from the entrance to the exit. In both experiments the left panel contains a long, winding path from entrance to exit, although in the second experiment there are a few loops and branches into culs-de-sac. The second fitness function in Experiment 1 also maximized path length. Since the agent that can jump 2 m can traverse the maze accessible to the agent that can jump 1 m, there is the potential for evolution

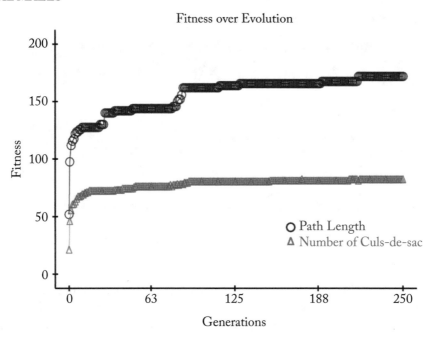

Figure 3.4: Fitnesses for a single run in Experiment 2. The upper plot gives the path length, the lower one counts the number of culs-de-sac in the maze. The selection fitness for the run is the geometric mean of these two values.

to select a maze in which both agents see the same maze. This did not happen. Figure 3.6 shows extreme examples of the behavior of the two fitness functions: wide variation and near convergence. Even at their closest approach, the lengths of the paths visible to the two different types of agents stayed 10 units apart. In most of the runs, they exhibited greater separation. This means that the "only one maze" optima of the fitness landscape are difficult to locate. This is especially interesting given the results in Experiment 3, which used the same pair of fitness functions on the generative representation. This experiment did have solutions where the two mazes were the same. It is worth noting that a great deal of additional area, off of the shortest path from entrance to exit, is accessible to the agent that can jump 2 m. Some of this consists of overlap with parts of the shortest path for the agent that jumps 1 m, but much of it is the result of genetic information not currently participating in the fitness-generating structures within the dual maze.

In Experiment 2 the fitness function for the agent that jumps 2 m was maximizing the number of culs-de-sac in that agent's embedded maze. As Figure 3.5 demonstrates, the character of the mazes visible to the two-meter-jumping agent is very different between Experiments 1 and 2. First of all, as one might expect, there are far more culs-de-sac. It is worth noting that

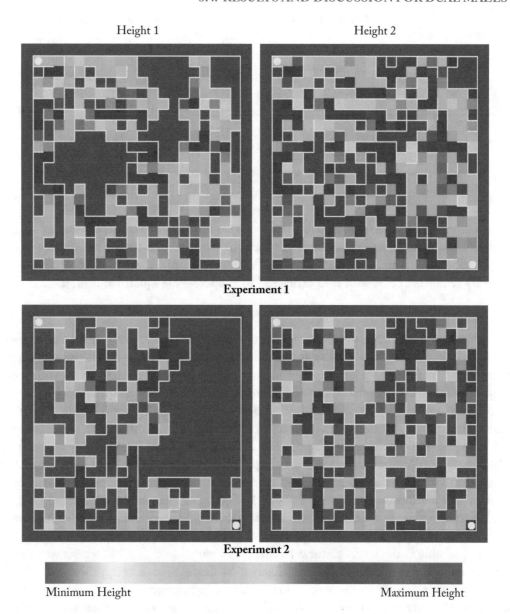

Height 1   Height 2

Experiment 1

Experiment 2

Minimum Height   Maximum Height

Figure 3.5: Example results from Experiments 1 and 2. The mazes on the left are those composed of squares accessible to the agent that can jump up or down one meter. The mazes on the right are the analogous set of squares for agents that can jump two meters. White lines are placed between adjacent squares where jumps are impossible. Yellow disks indicate the entrance and exit.

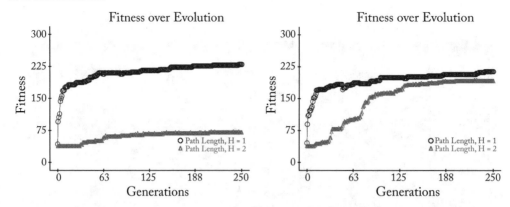

Figure 3.6: Plots of fitness over the course of evolution for two different replicates from Experiment 1. These plots represent extremes in which the path-lengths for the agents able to jump 1 m and to jump 2 m are either very different (left) or very similar (right). Colors between the maps vary somewhat because of JPG artifacts, but the accessible squares are correct.

some of the culs-de-sac carry over into the part of the map accessible to the agent that can jump one meter. The amount of inaccessible (gray) space for the 2-m-jumping agent also substantially decreases in Experiment 2 relative to Experiment 1. In a game where the agents with greater ability to jump were antagonists this could afford them much greater tactical flexibility.

Example outcomes from Experiments 3–6 are shown in Figure 3.7. These examples demonstrate that the outcomes change substantially when the fitness function changes, in agreement with the results from [13]. Experiment 3, like Experiment 1, uses f_{path} as the fitness for both embedded mazes. This means that Experiment 3 could also achieve high fitness from making the two mazes the same. Unlike Experiment 1, this actually happens in many of the replicates. The maze from Experiment 3 shown in Figure 3.7 has a very small number of its obstructions made of fire or water, and there is not much of a difference between the fitness values for the two agent types. This demonstrates that the behavior of the two representations is quite different, probably at the level of fitness landscape. A mutation that transforms a water or fire obstruction into a stone obstruction is very likely to be fitness neutral or improve fitness. Such a change cannot block paths from entrance to exit. Changing a height in the direct representation changes connectivity in a manner that is much less predictable and always has the potential to block the paths from entrance to exit. It seems likely that the difference in representation is responsible for the differing characters of the outcomes in these two experiments with the same fitness function.

In addition to the examples from Experiment 4 in Figure 3.7, Figure 3.1 also shows a result from Experiment 4. The fitness functions in this experiment emphasize path length for the fire-stone maze and culs-de-sac for the stone-water maze. The majority of the fire barriers

Experiment 3   Experiment 4

Experiment 5   Experiment 6

Figure 3.7: **Example results from Experiments 3, 4, 5, and 6. Green dots indicate checkpoints, while the small red circles mark the entrance and exit.**

serve to increase the path length of the maze, while the water is arranged to create a number of culs-de-sac. The two fitness functions f_{path} and f_{cul} are thus both achieving some degree of satisfaction. Interesting features that arise include a room (enclosed area) accessible only by an agent that can traverse the fire barrier in Figures 3.1 and 3.7, three similar rooms for an agent that can traverse water in Figure 3.1 and two in Figure 3.7, and two rooms in Figure 3.7 that have one wall made of fire and one made of water. This suggests that the issue of the degree of transparency of the water and fire barriers to each agent type may have substantial tactical consequences. A transparent water barrier might simply be a deep pool, while an opaque one might incorporate fountains or waterfalls.

Experiments 5 and 6 have fitness functions with very similar definitions. In both, the fitness function for the maze made of stone and water is to maximize the number of culs-de-sac. The stone and fire mazes have different fitness functions: f_{prim} and $f_{isoprim}$, both of which reward reconvergence after passing two checkpoints, on shortest paths from the entrance past both those checkpoints. The second of these functions differs from the first in that the reward is only granted if no other checkpoint is on a shortest path to the point that is a witness to the reconvergence. This second function is much harder to satisfy and, unlike all the others, is not helped much by using sparse initialization. As a result, the dynamics of Experiment 6 are driven by the need to have unique path reconvergence in the embedded maze defined by stone and fire barriers. Since the fitness used in selection is

$$\sqrt{f_{isoprim} \times f_{cul}},$$

if $f_{isoprim}$ is zero, the selection fitness is also zero. This causes selection to strongly favor obtaining at least one (of six possible, there are four checkpoints) nonzero isolated reconvergences. One such reconvergence is the most common outcome. Figure 3.8 shows fitness plots from Experiment 6 for a replicate in which the best gene has one nonzero isolated reconvergence (left) and a replicate that has three (right). Of the 30 replicates performed, 26 found only a single nonzero isolated reconvergence. Once more than one nonzero isolated reconvergence is located, the value of $f_{isoprim}$ in the population becomes much more variable because the space of mutations that do not yield a large drop in fitness is larger. Having a single isolated reconvergence forms a local optimum in the fitness landscape, one that is difficult to escape.

Because f_{prim} is much easier to satisfy, Experiment 5 is better behaved, with both fitness functions increasing rather than exhibiting large flat spots like those in the left panel of Figure 3.8. One feature of Experiment 5 is that it has an excellent chance of creating rooms (enclosed regions) accessible to only one of the two agents as well as creating single linear structures composed of multiple materials (as opposed to adjacent walls of different types). These types of structure are apparent in the example outcome from Experiment 5 shown in Figure 3.7.

The choice to simply amalgamate the two fitness functions via the geometric mean, rather than perform multicriteria optimization, was made for the sake of simplicity. In Experiments 1–5 this did not cause a problem. In Experiment 6 the fact that the $f_{isoprim}$ function was difficult

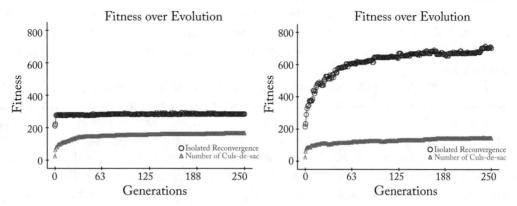

Figure 3.8: Two plots of fitness over the course of evolution from Experiment 6. These plots represent extremes. On the left all fitness from $f_{isoprim}$ results from the isolated reconvergence of a single pair of checkpoints, while the right hand side benefits from nonzero isolated reconvergence from three pairs of checkpoints.

to optimize and was capable of taking on a zero value meant the geometric mean choice was a poor one which caused unbalanced outcomes: overuse of fire. The obvious conclusion is that multicriteria optimization should be implemented. This has the added advantage that multiple fitness functions could be optimized for a single maze as well.

3.5   CONCLUSIONS AND NEXT STEPS FOR DUAL MAZES

The primary difference between the techniques presented in this chapter and Chapter 2 and those in Chapter 1, in comparison to other procedural content generation techniques, is the use of checkpoint driven dynamic programming to obtain a far larger set of elements for designing fitness functions and so obtain finer control over the types of mazes produced by the automatic system. This is also in contrast to the current industry standard of hand design by teams of skilled programmers and artists. The primary advantage of automation is speed; the work presented in Chapters 2 and 3 increases the control a game designer has over that speed. They provide a demonstration of the control a game designer can have over the content, via the choice of fitness functions, while maintaining the increase of speed from partial automation.

Chapter 3 also demonstrates that the techniques introduced in earlier chapters to design dynamic programming based fitness functions can be generalized to dual mazes. Experiment 6 serves as a warning that a shift to more sophisticated techniques for simultaneously satisfying multiple fitness functions are required, probably some form of multicriteria optimization. Two different representations for dual mazes were tested. The first is a direct representation that directly specifies the heights of a terrain map. The dual nature of the maze is realized by assuming two types of agents that have different maximum heights they can jump. The second representation is generative and places linear barriers made of stone, fire, or water on an originally empty grid. The embedded mazes defined by this representation are the maze whose walls are fire or stone and the maze whose walls are water or stone. The dual maze structure is easier to visualize in this representation.

The evolutionary experiments met the goal of automatically designing dual mazes. In Experiment 4 mazes were evolved that had a very long path length in the embedded fire and stone maze and relatively high connectivity in the embedded water and stone mazes. Both fitness functions were, to a substantial degree, satisfied. The different representations were shown to react very differently to both mazes having the same fitness function, as shown by the contrast between Experiments 1 and 3. Both experiments attempted to maximize the path length from entrance to exit for both mazes. In the direct height-based representation the two mazes were usually very different, while the generative representation was able to evolve two very similar mazes. This latter state is an optimal solution if the goal is maximizing fitness, but was far harder to reach when using the direct representation.

A total of four representations were used in Chapter 2 with an assortment of fitness functions. One of these representations was modified, by allowing multiple barrier types, to yield the generative representation used in this section. Another, the *chromatic mazes*, was modified, replacing colors with real-valued heights to obtain the direct representation used in this study. Four novel fitness functions, combining elements used in the earlier studies, appeared here. This demonstrates that the creation of new representations and fitness functions within this domain is *easy*. Figure 3.9 shows the results of modifying the generative representation to disallow diagonal obstructions. The resulting mazes are very different, but the time required to make the modification to the code was minutes—and reversible in seconds. A maze, using the setup of Experiment 4 with different grid dimension, also appears in Figure 3.9. Both these serve as examples of the flexibility of the technique.

3.5.1   ADDITIONAL FITNESS ELEMENTS

There are a large number of possible elements that could be added to the list of numerical quantities used to create fitness functions.

- One easily computed quantity that is relevant to the character of the maze is the number of inaccessible squares. Minimizing or maximizing this quantity, in the presence of other constraints, could be used to grant added control over the type of maze that evolves.

- At present, the shortest path distance between checkpoints or checkpoints and the entrance and exit is not used. Maximizing or minimizing these in various combinations would permit a designer to pre-specify an important tactical element but leave a great deal of room for evolution to create different mazes that satisfy those constraints.

- *Field of view* from a designated set of points within the maze is a rich area for future exploration. In a maze with a single type of barrier one could ask that the part of the maze

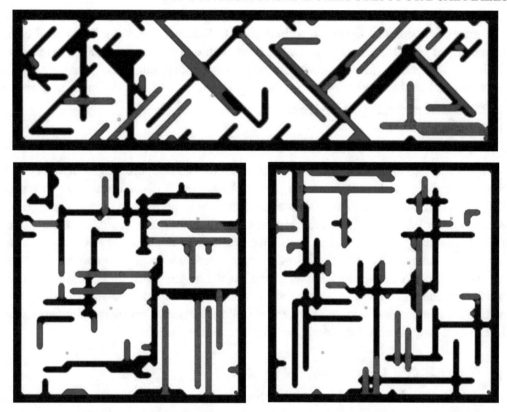

Figure 3.9: An example of Experiment 4 rerun on an 80 × 20 grid (note the checkpoints were moved) and two examples of the results of rerunning Experiment 4 with the ability to have diagonal walls removed. Red disks denote the entrance and exit, while green disks denote checkpoints.

visible from a small set of points, where the designer might put guards or monsters, be maximized. By itself this fitness function has an obvious global maximum: no barriers. In conjunction with other fitness functions, visibility maximization might well yield interesting structures. In a dual maze, if different types of agents experienced different levels of transparency for different types of barriers or levels of illumination, one could optimize the ratio of viewpoints. This would permit a designer to have strong control over the tactical parameters of a maze.

- Another type of fitness element that could be incorporated is the behavior of simple agents introduced into a maze. A virtual robot's ability to encounter checkpoints or move from the entrance to the exit of a maze would serve as another means of evaluating a maze.

- It is not difficult to get a dynamic programming algorithm to generate the *number of paths* between one point in the maze and another. Calculating these numbers would enlarge the design space for fitness functions. A maze with one path from entrance to exit is very different from one with dozens. Optimizing the ratio of the number of paths reaching one checkpoint to the number of paths reaching another would bias the maze to have more connectivity in one part of the maze.

3.5.2   TOOL DEVELOPMENT

An obvious next step for this research is to build a GUI tool for designing maze-like game levels. In this context the algorithm would act as a designer's assistant. A dialog box for piecing together a fitness function out of the elements described here would permit a designer a great deal of freedom to control the types of maze that evolve, after which the mazes could be transferred to an editor for polishing. Figure 3.3 shows, in its lower panels, a re-rendering of one of the outcomes that make the connectivity of the embedded mazes more obvious. This is another example of a key map of the sort defined in Chapter 1, but for the height-based direct representation. If the quantities computed by the dynamic programming algorithm are available to a design tool, then multiple renderings of a maze, including planning representations that are not available to a player, could be quite useful to a designer and enhance their productivity.

3.5.3   VISIBILITY AND LINES OF SIGHT

The issue of *transparency of barriers*, raised in Section 2.7, yields another direction for this research. Imagine that a wall of fire can be either easy to see through or smoky and hence opaque. Likewise, a water barrier might be a deep pool or a waterfall with a great deal of spray. In this situation, rather than stone, fire, and water, we might have stone, clear fire, smoky fire, deep pools, and waterfalls. Use of a representation with variable transparency would only be useful if variables concerning line-of-sight were used in the fitness function.

Another potential factor that could be incorporated is level of illumination. Agent types might have no night vision, adequate night vision, or perfect darksight. With these different illumination types, embedded mazes could be defined by level of illumination. At this point the maze might be the streets of a city, with a representation that places buildings. A sparse, safe (well-lit) network of major streets could form one embedded maze, while a poorly lit but much larger embedded maze could comprise the back alleys and abandoned lots.

3.5.4   TERRAIN TYPES

A natural application of embedded mazes would be the creation of terrain maps with different terrain types. In a military or resource-economic simulation the useful terrain types might be water and several land types defined as passable by wheeled vehicles, tracked vehicles, horses, infantry, or as completely impassable. The embedded mazes would then represent areas accessible to different types of units. For this type of terrain generation, multicriteria optimization would almost certainly be required.

CHAPTER 4

Terrain Maps

Landscape induction, the creation of landscapes with desired properties, is an active area of research in automatic content generation for games. A common way of specifying a landscape or terrain map is with a *height map*, a specification of the height of the landscape at each point on a grid. The dynamic programming based techniques in the previous chapters can be used to stitch together terrain features, and so this chapter will concentrate on two different representations for evolving terrain features in tiles.

While we will return to a dynamic programming-based fitness function at the end of the chapter, the primary fitness function used to evolve landscape features is one that measures the degree to which the evolved landscape matches an idealized landscape. It is easy to make a perfectly, symmetrical landscape with mathematics. Evolving approximations to an idealized landscape yields a variety of plausible landscape features. In addition, the representations given in this chapter are *multi-scale* in the sense that they can be rendered at various resolutions to give plausible images at different apparent viewpoint distances.

Three examples of idealized landscape features are given in Figure 4.1. The fitness function used, for each landscape, is the root-mean-squared error (RMS error) between the height map generated by the evolving height map generator and the idealized landscape. The formula for this fitness function appears in Equation (4.1). Note that, unlike previous examples, this fitness function is to be *minimized*:

$$RMSE = \sqrt{\frac{\sum\limits_{p_n} (H_e(p_n) - f(p_n))^2}{n}}, \tag{4.1}$$

where the p_n are the points in the sample grid; H_e is the value of the evolved height; and f defines the idealized landscape.

The various functions shown in Figure 4.1 are all generated with the same trick. If $f(x, y) = c$ defines a curve in the plane, then the formula given in Equation (4.2) graphs as a surface with a ridge of height one over the graph of $f(x, y) = c$, dropping away from the ridge toward zero. This technique can be used to generate a large variety of idealized, simple landscapes:

$$height(x, y) = \frac{1}{(f(x, y) - c)^2 + 1}. \tag{4.2}$$

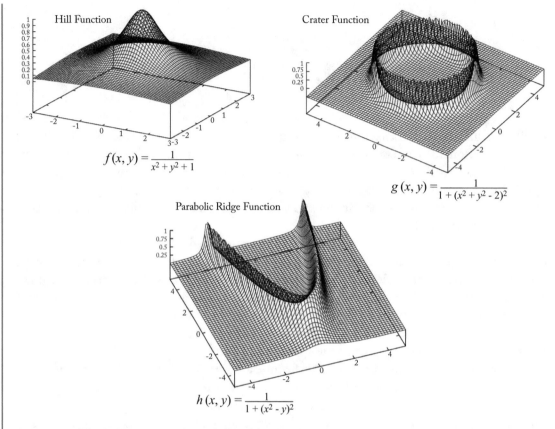

$$f(x, y) = \frac{1}{x^2 + y^2 + 1}$$

$$g(x, y) = \frac{1}{1 + (x^2 + y^2 - 2)^2}$$

$$h(x, y) = \frac{1}{1 + (x^2 - y^2)^2}$$

Figure 4.1: The hill, crater, and parabolic ridge landscapes.

4.1 MIDPOINT L-SYSTEMS

An L-system or Lindenmayer system [47, 58] consists of two parts. The first is a grammar which specifies an axiom and a collection of replacement rules. The L-system creates a sequence of objects, starting with the axiom, by applying the rules. This application is called an *expansion* of the L-system. In this study the replacement rules operate on a two-dimensional array of characters. Rules are applied simultaneously to every symbol in a two-dimensional array to create the next array. An example of this type of two-dimensional L-system is shown in Figure 4.2.

The second part of an L-system is the *interpreter*. The interpreter's task is to render the symbols into the desired type of object, in this case the height map that generates a virtual landscape. In midpoint L-systems interpretation is integrated with expansion using the following rules.

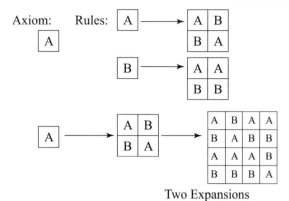

Two Expansions

Figure 4.2: An example of a midpoint L-system.

1. In each expansion of the system, each grid square has a *height* associated with it. The height of the axiom is zero and, at the end of expansion, the heights of the grid squares will be the entries of the height map.

2. In each expansion of the system, each grid square has a *decay factor* associated with it. This is used, as a multiplier, to reduce subsequent changes in height from every grid square that arises from the grid square. The decay factor of the axiom is 1.0.

3. Each symbol has a Δ_h value that represents the change in height when that symbol is applied. This value is modified by a the current decay factor and then added to the grid square's height when the grid square is first created by an expansion.

4. Each symbol has an α value that represents the decay of height increase when that symbol is applied. After the current decay factor is applied to the current change in height, α is multiplied by the grid square's current decay factor to get a new decay factor.

5. When the rule associated with a grid square is expanded, its current height and decay factor are passed to all the daughter grids.

The *decay factor* was a feature added to this system during testing. Without the decay factor, the resulting landscapes are choppy and chaotic. Having changes in height decay with the depth of the expansion permits much smoother landscapes to be evolved.

Algorithm 1 shows the algorithm used to perform expression of a midpoint L-system. Since testing was performed with relatively small grids of height values, this algorithm was more than fast enough. Additional speed could be secured by adding a test for the decay factor becoming small enough that there is no reason not to prune the tree.

Algorithm 1 Midpoint Lsystem Expression
 Input: *L-system rules, final depth*
Output: *A height map for each small grid*
Details:

Initialize decay=1.0
Initialize height=0.0
Initialize depth=0
Call Traverse(initial grid,height,decay,depth)
Report Heights.

Procedure Traverse(grid,height,decay,depth)
   if(depth=final depth)return(height) else
      Given the symbol in the current grid
      For each subgrid
         decay_new=decayα*
         height_new=height+$\Delta_h \times decay$
         Traverse(subgrid,new_height,new_decay,depth+1)
      End else
   End For
End Procedure Traverse

4.1.1   THE REPRESENTATION FOR MIDPOINT L-SYSTEMS

The fundamental feature of a midpoint L-system is the number of rules it uses. These rules are designated with capital letters, like the "A" and "B" in Figure 4.2. For each symbol, the four symbols it is expanded into must be specified as well as the numerical parameters Δ_h and α for change in height and increase in decay. An example of a midpoint L-system evolved to produce a hill function is shown in Figure 4.3. The symbols are mapped onto the 2×2 grid for expansion in reading order.

In an initial population the values that define an instance of the representation are filled in uniformly at random. The height changes Δ_h are chosen uniformly at random in the range $[-1, 1]$, while the decay factors changes in the initial population are chosen in the range $[0.5, 1.0]$. The four symbols and Δ_h and α values that define the behavior of a symbol in the midpoint L-system are treated as atomic object for crossover. The crossover takes place on the linear structure of these symbols. The crossover operator used is *two-point crossover*, which exchanges middle segments of the list of symbol behavior descriptions with the middle segments selected uniformly at random. The mutation operator replaces one symbol or numerical value with one chosen uniformly at random.

Rule	Expansion	Δ_h	α
A	EEFC	-0.0932	0.508
B	BBBD	-0.0973	0.513
C	CBBD	0.446	0.992
D	AABB	-0.223	0.500
E	EEFF	0.0479	0.762
F	DFBB	0.533	0.500

Figure 4.3: An example of a six-symbol midpoint L-system.

The evolutionary algorithm used is the same as the one used in Chapter 2 except for the change in the representation used for members of the evolving population and the use of an RMS fitness function. A parameter-setting experiment was run with the hill function testing different population sizes and numbers of symbols. The numbers of symbols used were 6, 12, and 18, and the population sizes used were 10, 32, 100, 320, and 1,000, representing roughly equal multiplicative spacing. For each of the 15 pairs of parameters tested the evolutionary algorithm was run 30 times. The results of this parameter study are shown in Figure 4.4.

The parameter setting experiment shows that using more symbols produces substantially better results. In addition, while it is a smaller effect, larger populations yield better results. The populations of size 320 and 1,000 yield similar results, suggesting that populations of size 1,000 saturate the system's need for initial information content. In the 18-symbol experiments, the performance of the 320 member populations was superior.

If more symbols yield better performance, a natural question is why one would not use a very large number of symbols. As the number of symbols increases the asymptotic performance of the L-system in modeling an idealized land-form cannot get worse, but it may not get much better. Each symbol's behavior requires four symbols and two real numbers to specify. This means that increasing the number of symbols substantially increases the amount of information that the evolutionary algorithm must discover. This means that the smallest number of symbols that provide acceptable performance yields the most efficient configuration of the system.

4.1.2   MULTISCALE LANDFORMS

One of the nice features of using the midpoint L-system representation is that it grows the hills in a stepwise fashion. A common problem in the rendering of landscapes is dealing with objects that are nearby as opposed to far away. Each expansion of a midpoint L-system yields an additional level of detail. Figure 4.5 shows the same L-system expanded 4, 5, 6, and 7 times. With each expansion the hill becomes more detailed.

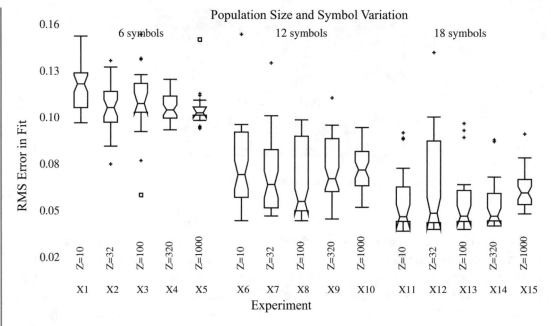

Figure 4.4: Results of parameter setting on the hill function. The parameter Z is the population size. Shown are wasp-plots for the best final fitness in 30 runs of the evolutionary algorithm for each set of parameters tested. Fitness is RMS error with the hill function, which is better when it is smaller.

Figure 4.6 shows six different evolved craters, expanded six times. These craters are a bit blocky and increasing the evolution time or applying a simple spatial smoothing model to them will yield less choppy craters. The last two craters in Figure 4.6 are the first and sixth run through the filter given in Equation (4.3). This equation averages a 3 × 3 grid of heights giving the center value four times the weight of the others:

$$h_{i,j}^{new} = \frac{1}{12}\left(\left(\sum_{k=-1}^{1}\sum_{m=-1}^{1} h_{i+k,j+m}\right) + 3h_{i,j}\right). \tag{4.3}$$

One of the main reasons for evolving the midpoint L-systems is to grant a digital artist a variety of different forms of hill, crater, or other landscape. These objects form a palette of basic land forms that can then be stitched together. Some normalization may be needed to create smooth, flowing landscapes, but this can be done in a fashion that is not computationally costly.

The multi-scale feature of the L-system based landscape features has another interesting possibility embedded in it. Each grid square in the L-system could, itself, be treated as an axiom for another L-system instead of the next call in the current L-system. This means that calls to a set of land forms, hills, and craters could become characters in an L-system. Most of the squares

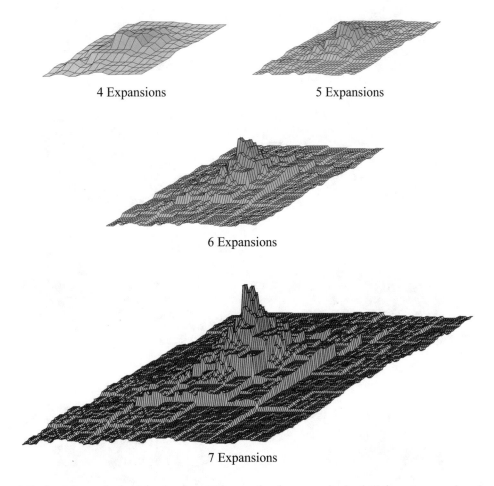

4 Expansions   5 Expansions

6 Expansions

7 Expansions

Figure 4.5: An 18-symbol midpoint L-system evolved to match the hill function rendered at 4 different resolutions.

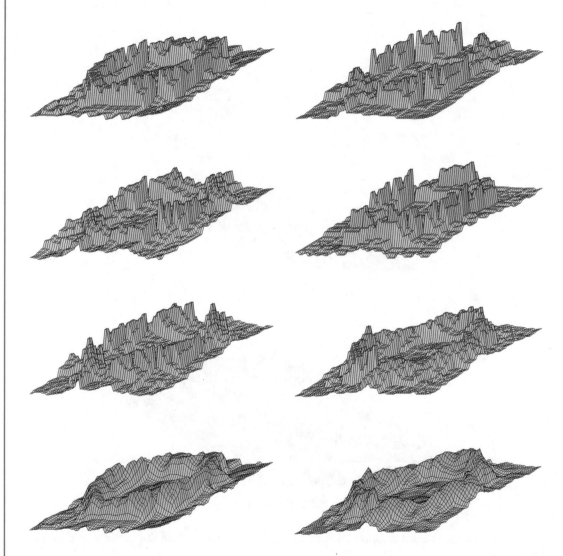

Figure 4.6: Examples of different evolved 24-symbol craters shown at 6 expansions. The bottom two craters have height maps that have been subjected to averaging to smooth the landscape.

in an L-system expand to sets of rules in that L-system, but some expansions could transfer control, in their grid square, to another L-system. This means that an entire, complex landscape could be specified by a single master L-system.

When rendering this sort of master L-system, a level of depth and a spatial coordinate could be used to query the L-system for a height at a specific point on a specific grid square. It is worth noting that it takes 16 bytes to specify the 4 symbols and 2 real numbers that specify the behavior of a symbol in a midpoint L-system. This means that an 18-symbol midpoint L-system requires 288 bytes to specify. This sparsity of description permits thousands of L-system in multiple layers to be used to describe a large landscape in very little space.

On the other hand, evolving an entire landscape in the fashion described would be a nightmare problem. The problem, however, is eminently decomposable. The craters and hills—and other land forms that will appear in the next section—can be evolved and banked and then other L-systems with different goals, such as a particular tactical connectivity like those studied in Chapter 2, can be used at a higher level. This places very small data specifiers for a complex landscape well within reach.

In the next section, we will also see that, by adopting the connectivity rules of the height maps from Chapter 3 and checkpoints, we can evolve landscape maps that have a specified connectivity.

Both midpoint L-system and the landscape automata featured in the next section generate height maps, that is to say a rectangular grid with heights specified. The algorithm for rendering these into a map is shown in Algorithm 2

Algorithm 2 Rendering a height map

Input: *A height map and a viewpoint in space.*
Output: *a picture of the encoded landscape..*
Details:

Treat the entries of the grid as points (x,y,height)
Let $\vec{v}$ be the vector from the viewpoint to the grid's center
Project the points (x,y,z) onto the plane orthogonal to $\vec{v}$
Make quadrilaterals of points from groups forming a square in the grid.
Sort the quadrilaterals by distance of the center from the viewpoint.
From farthest to nearest center of mass for quadrilaterals
  Set gray scale based on height of center of mass
  Draw and fill the quadrilateral
End From

The correct landscape perspective arises from filling the quadrilaterals from farthest to nearest, a trick called *Z-buffering*. Normally, this sort of content would hand off the quadrilaterals to a rendering engine such as Open-GL or MESA.

4.2   LANDSCAPE AUTOMATA: ANOTHER REPRESENTATION FOR HEIGHT MAPS

This section introduces a new representation, called *landscape automata* (LSA), for searching the space of height maps. Search is performed with an evolutionary algorithm and so is another example of SBPCG, a variant of PCG which incorporates search rather than trying to write a PCG system that can generate acceptable content in a single pass. In this section, two different methods of granting a designer control over the height maps generated by LSAs are demonstrated. The first is the one used on midpoint automata; it permits the designer to specify an idealized landform, like a hill, crater, or ridge which an evolutionary algorithm then approximates. Both good and mediocre approximations result, providing a palette of landforms for the designer. The second control method treats the height map as a maze in which barriers consist of height differentials between adjacent grid squares of the height map that exceed a critical threshold. This is similar to the height maps from Chapter 3. This method permits the designer to specify checkpoints that must be mutually accessible and permits the designer to then use any of several fitness functions based on characteristics of paths between fitness functions. Many fitness functions of this sort are given in [15]; in this study four checkpoints are used, and the algorithm maximizes the total pairwise distance between them.

The work in the current study grows out of work published in both [26] and [15]. The second of these demonstrated four representations for encoding evolvable mazes: a direct encoding, a positive and a negative generative representation, and a representation as a height map. In this study we demonstrate that LSA can both approximately match idealized terrain models to provide a palette of terrain features and also create mazes specified by a height map and a threshold value.

4.2.1   DEFINING LANDSCAPE AUTOMATA

Landscape automata are a type of self-driving finite state device that repeatedly partitions a height map down to the pixel level while simultaneously generating a height map. Partitioning is accompanied by state transitions driven by the identity (quadrant number) of the partition region. Each state contains a height modifier, to be added to the height, when a partitioning yields a transition to the height and a cascading decay parameter that acts to limit the increase in height as more transitions are made.

Figure 4.7 shows the partitioning process. The regions in a given partitioning are numbered 0, 1, 2, 3 in reading order:

0	1
2	3

An example of a LSA is given in Figure 4.8. To generate a height map the automata is called at state 0 with an initial height of zero and an initial height multiplier of 1.0. The drawing area is then partitioned into quadrants and the automata recursively calls itself in each quadrant, making the appropriate transition for that quadrant. Each chain of downward transitions is separate, maintaining its own local values for both height and the accumulated decay parameter. It might be profitable to view the recursive calls to a landscape automata as being structured by a quadtree [66]. Quadtrees are trees in which each branch is a four-quadrant partition of an image. As the automata enters a state it updates the recursive multiplier value by multiplying it by the decay parameter associated with the state. The automata then adds its height value, e.g., "+4", multiplied by the recursive decay parameter, to the height along this chain of calls. The recursive calls terminate when the drawing area is the size of a single grid square of the final height map, in which case the current height value is simply assigned to that grid square. A

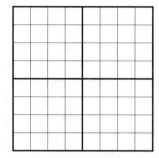

Figure 4.7: Recursive quadrature of the map region.

| | Transitions | | | | Height |
	If(0)	If(1)	If(2)	If(3)	Multiplier
S0	+3 → S1	+2 → S2	+1 → S3	+1 → S0	0.995
S1	+4 → S0	+2 → S0	+1 → S0	+3 → S0	0.876
S2	+1 → S3	+3 → S1	+0 → S1	+2 → S2	0.788
S3	+0 → S3	+4 → S2	+4 → S3	+2 → S3	0.964

Figure 4.8: A landscape automata with four states. States have four transitions, each of which is composed of a height adder and a transition for each of the four quadrants together with an overall height multiplier for the state.

LSA thus specifies a height map for a given size of square drawing arena, filling a square array with height values. Notice that an LSA makes no use of random numbers—it is a deterministic specification of a height map. LSAs share with midpoint L-systems the property that they can be called at less than their full resolution to generate a coarser height map suitable for distant views of the object encoded.

Crossover of LSA is performed by treating the list of states as a linear chromosome and subjecting it to two-point crossover. A point mutation is accomplished by first picking a state and then picking one of the nine objects (four height adders, four transitions, one decay parameter) and modifying one of them. Height adders and transitions are selected uniformly at random. The state's height multiplier is modified by adding a uniformly distributed number in the range $[-0.1, 0.1]$, and then reflecting any values that wander outside of $[0, 1]$ across the appropriate boundary, either $x \to -x$ or $x \to 2 - x$. Height adders take on integer values $\{0, 1, 2, 3, 4\}$ both during initial population generation and mutation. The height multipliers are initialized in the range $[0.75, 1.0]$ with the value selected uniformly at random.

4.2.2   EXPERIMENTS WITH LANDSCAPE AUTOMATA

The evolutionary algorithm used to evolve landscape automata is steady state [67] operating on a population of 100 LSA. Selection and replacement are performed with size four single tournament selection. This model of evolution selects four members of the population, copying the two better over the two worst. The list of states of the two copies are subject to two-point crossover. Each copy is then subject to $1 - N$ mutations, with the number chosen uniformly at random, for $N = 4, 7, 11$, and 15. Two types of fitness functions are used. The first takes a specialized idealized landscape feature and attempts to minimize RMS error of the normalized height map generated by the LSA with the landform. This is the same fitness function used on L-systems. The second fitness function uses a dynamic programming based function to maximize the total distance between each pair of a check points placed in the terrain grid. Evolution is continued for 40,000 mating events and each experiment is comprised of 30 independent replicates of the evolutionary algorithm.

The second type of fitness function treats the height map as a maze in a manner similar to that done in Chapter 3 where passage between two grid squares of the height map is possible if their difference in height is less than a critical value. This study uses the critical height $H_{crit} = 0.1$. Recalling that the height is restricted to lie in a relatively small range, this is a fairly steep value: it makes the maze more visible in figures. Four checkpoints, one in the center of each edge of the height map are used. The six pairs of distances between these checkpoints are computed, and the evolutionary algorithm uses their total as its objective function. If the four checkpoints are not all mutually accessible, the LSA that generated the height map is assigned a fitness of zero. This objective function creates a maze with long, winding passages joining the checkpoints. This objective function is called the **maze** function. Note that, unlike RMS error, the maze function is *maximized*.

4.2.3   RESULTS AND DISCUSSION FOR LANDSCAPE AUTOMATA

The number of states in an LSA have a significant impact on the quality of solutions. Box plots comparing 10, 30, and 90 state automata for the hill function are shown in Figure 4.9.

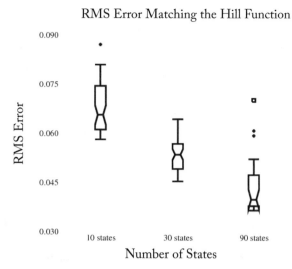

Figure 4.9: Impact on solution quality of varying the number of states in an LSA for the hill function. Recall that the fitness function in these experiments is *minimized*.

Figure 4.10 shows the impact of varying the distribution of the number of mutations when the evolutionary algorithm is being matched to the hill and crater functions, respectively. In both cases, fewer mutations are uniformly better, but the effect in not large. The comparison between the smallest and largest ranges for numbers of mutations is significant in both cases. The best choice for both the crater and hill functions is thus 90 states and 1–4 mutations; this value was adopted without testing for the parabolic ridge.

Figure 4.11 shows the impact of changing the number of states in the LSAs for the maze fitness function. The result is different from that for the landform matching experiments. The best average fitness results from using 10 states, but the highest fitness individual was located in the 90-state experiment. This suggests that the populations were still improving with 90-state LSAs requiring more evolution to reach their full potential.

The different results, that many states are good for landscape matching and fewer are better for making a height map that yields a long and winding maze, show that the number of states and the problem being solved interact in a non-trivial way. A map that operates by iterated vertical displacement will find the simulation of curved surfaces challenging; giving it more states grants it more degrees of freedom to deal with the problem. For a long and winding maze the problem is simpler. This is not obvious. It is an experimental result, deduced from Figures 4.9 and 4.11.

Figure 4.12 shows the same evolved height map rendered as both a terrain map and as a maze. The maze rendering, while hiding much detail, directly exhibits possible and impossible moves between terrain grids. The use of such dual views returns us to the discussion of *key maps*

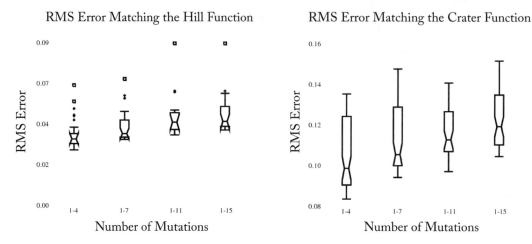

Figure 4.10: Impact on solution quality of varying the range of number of mutation for LSA matching the hill function (left) and the crater function (right). Recall that the fitness function in these experiments is *minimized*.

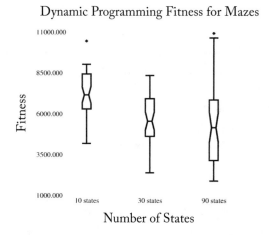

Figure 4.11: Impact on solution quality of varying the number of states in an LSA for the maze fitness function. Recall that this function is *maximized*.

from Chapter 1. As before, these multiple views have the potential to grant a game designer power in the form of an omniscient abstraction (the maze), while giving the player a much less lucid, but more realistic, view. Notice the correspondence of features; plateaus in the landscape correspond to plazas in the maze.

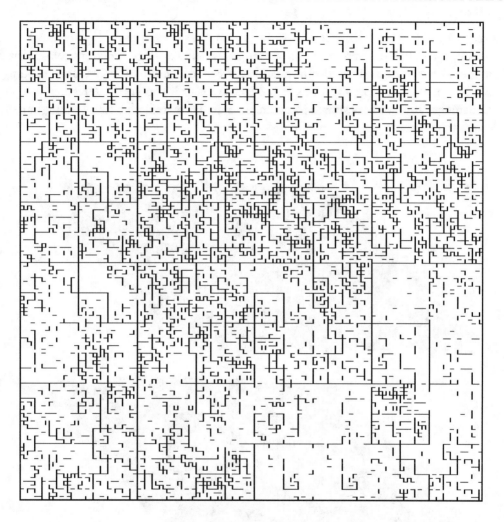

Figure 4.12: An evolved landscape displayed as a maze—walls represent boundaries too steep to climb—and as an explicit terrain map.

4.2.4 QUALITATIVE DIVERSITY

Figures 4.13, 4.14, 4.15, 4.16, and 4.17 show examples of terrain renderings of height maps for the hill, three different version of the crater, and parabolic ridge functions. Note that the different examples for the same idealized landform can look quite different. The encoding as an LSA can yield good approximations to the landforms or it can yield interesting, but rough approximations. The two hills shown in Figure 4.13 are both creditable hills but might benefit from smoothing.

Figure 4.13: Two examples of terrain renderings of LSA evolved to match the hill function.

The craters shown in Figure 4.14 are not bad, although the first has a gap in the crater wall, facing the reader. The interior topography at the bottom of the crater also varies from example to example. Figure 4.15 shows examples of more rugose craters with ridges and peaks adjacent to the crater. The craters in Figure 4.16 are especially interesting. When making transitions, the landscape automata can transition to the starting state of the automata. Nothing in the representation forbids this behavior, other than the potential for poor fitness. When this happens

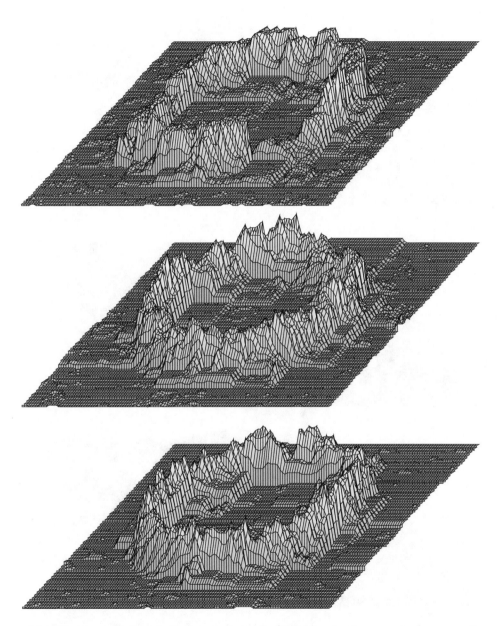

Figure 4.14: Three examples of terrain renderings of LSA evolved to match the crater function that yield relatively smooth craters.

Figure 4.15: Three examples of terrain renderings of LSA evolved to match the crater function that yield more rugose craters.

Figure 4.16: Three examples of terrain renderings of LSA evolved to match the crater function that evolved to exhibit multiple craters, suggesting a self-symmetric landscape automata.

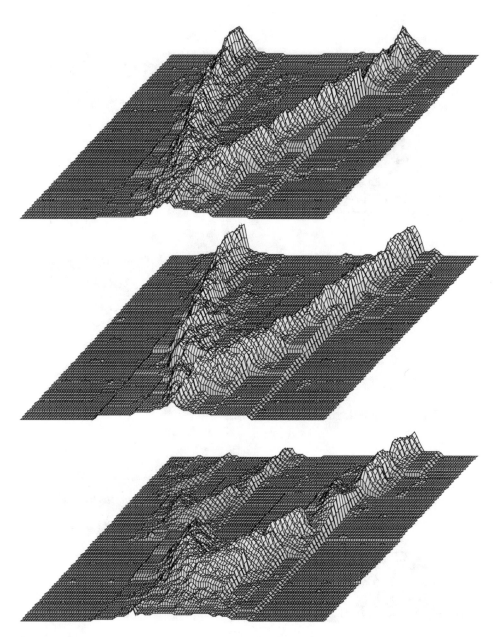

Figure 4.17: Three examples of terrain renderings of LSA evolved to match the parabolic ridge function. The third exhibits some self-symmetry.

it can yield multiple smaller craters or craters interrupting craters. This was an unanticipated and fortuitous property of this representation.

Figure 4.17 shows examples of parabolic ridge approximations. The third is 1 of 2, out of 30 trials, that has the property of looping back to its starting state. The fitness of a parabolic ridge is reduced far more by most ways a secondary ridge can be introduced into the landscape, and so this self-similar behavior is much rarer in the parabolic ridge landscape automata than those evolved to minimize craters.

The many variations of the theme of the original idealized landform generated by the evolution of LSAs forms a potential strength of the representation. It both gives the designer a compact form of the height map they asked for and also suggests intriguing variations of it. This is something that happens often in the area of evolved art—local optima of the fitness function are often of similar desirability to any global optima that are located. Since LSA can only approach continuous, curved surfaces in the limit, they must approximate them when a finite number of grid squares are used, a fact that drives diversity as many diverse local optima approximate the ideal curve.

4.2.5   CONCLUSIONS AND NEXT STEPS FOR LANDSCAPE AUTOMATA

This section has demonstrated that LSAs can be used as small objects that store variations on designer-specified landforms. They share this with midpoint L-systems but the two structures operate in a very different fashion. LSA also serve as a representation for height maps that can create terrain with designer-specified connectivity properties. This can be extended with techniques similar to those in [50] to place particular objects like towns or ammunition dumps into the landscape with desired relative distances. In [18] individual maps were used as tiles to efficiently design gigantic level maps, something we will explore in Chapter 5. These techniques can also be applied to the terrain maps generated with LSAs.

It may also be worth reiterating a type of scalability that LSAs share with midpoint L-systems. An LSA that is evolved with a fitness function using a 128×128 grid can be called on a smaller grid that it will fill with a lower-resolution version of the same object. This scaling property also suggests a way to speed up evolution: evolve initially with fewer sampling points. The accumulation of the decay parameter forces information about the landform to be encoded with larger features higher in the recursive call tree and details lower. This thought explains why LSAs have this scaling property.

Another property of LSAs not yet demonstrated is the ability to cascade LSAs. A hill could be studded with craters by simply passing the recursive call from an LSA that builds a hill to one that builds craters. This cascading property can be used to any depth, and the height in the quadtree decomposition of the landscape where the hand-off happens controls the size of the subsidiary features.

Next Steps for Landscape Automata

Single parent techniques [28] augment a standard evolutionary algorithm with an additional variation operator called *single parent crossover*. A selection of structures of the same type as the evolving population, called the *ancestor set* is added to the algorithm. Whatever crossover operator is used by the algorithm is used, in a one-sided fashion, by single parent crossover to permit population members to cross over with members of the ancestor set. This technique ensures that critical building blocks cannot be lost and provides domain knowledge to the degree that the ancestors are high quality solutions to the problem. While modest care is needed to ensure that the ancestors are not simply cloned via multiple crossover events, the technique can substantially reduce time to solution or improve solution quality.

In [20] it was shown that single parent crossover also has the ability to substantially focus evolutionary search in the vicinity of the ancestor set. This suggests that single parent techniques would be an excellent next step for LSA. Both improving solution quality and focusing an algorithm near an interesting variation on a landform are potentially desirable. Since they are finite state devices that can encode their functionality in a subset of their genome, LSAs are well suited to single parent crossover.

4.3   MORPHING AND SMOOTHING OF HEIGHT MAPS

A concluding point worthy of emphasis follows from the two smoother created in Figure 4.6. A midpoint L-system, an LSA, or any other technique for generating a height map produces an array of heights. The crater simulations of the midpoint L-systems in Section 4.1 were too rugose. A simple averaging filter smoothed them out, yielding more plausible craters. This suggests that any landscape generator can be mathematically modified to produce a smoother landscape. So far this is obvious.

If two height maps are derived from the same idealized landscape then *their average is also an approximation of that landscape*. This thought can be applied in a number of ways.

- The weighted average of two evolved landscape features is another landscape feature. By varying the weight one can transform one of the landscape features into another.

- While an averaging filter smooths a land form, there is a better way to reduce the roughness. If the original idealized land form is available, a weighted average with it creates an entire space of landforms with the idealized land form at one point and the set of evolved landscape features at other points. A one-third ideal, two-thirds evolved crater may be exactly the crater that a designer wants.

- The sum of two landscape features is a landscape feature. This means that we could add several hills together, for example, to get a complex mountain. This is an alternative to making an idealized multiple hill and trying to match it.

CHAPTER 5

Cellular Automata Based Maps

5.1 FASHION-BASED CELLULAR AUTOMATA

This chapter builds on and extends earlier work using a cellular automaton to design level maps resembling a network of caverns that appear in [43]. The earlier work used a majority-based automata, a rule that does well at clumping full and empty cells to yield a plausible cavern map. A system for automatic content generation (ACG) of level maps based on a novel type of cellular automata is developed, tested, and variations are developed. Evolution is used to search the space of automata rules that find ones that yield maps with large open areas and plausible walls.

Cellular automata instantiate discrete models of computation. A cellular automaton has four components.

1. A collection of cells. In this work a grid of cells that make up the map being created.

2. A designation, for each cell, of which set of cells form the neighborhood of the cell. In this work this will be the cell itself and the four cells closest to it in a grid, as shown in Figure 5.1.

3. A set of states that cells can take on, in this case "empty" with a numerical value of zero and several colors of "full" with different numerical values.

4. An updating rule that maps the states of a cell's neighborhood in the current time step to a new state for the cell in the next time step.

Figure 5.1: A von Neumann neighborhood of the cell **C**. The members of the neighborhood are the cells are those numbered 1, 2, 3, 4.

In practice, cellular automata are a type of discrete dynamical system that exhibits self-organizing behavior. When a cell population is updated according to local transition rules, it can form complex patterns. The updating may be synchronous, as it is here, or asynchronous. Cellular automata are potentially valuable models for complex natural systems that contain large numbers of identical components experiencing local interactions [60, 73]. Examples of cellular automata based cavern maps are shown in Figure 5.2. The automata in this study are called *fashion-based* cellular automata because the updating rule may be thought of as following the current fashion within each neighborhood. The locality property of cellular automata turns out to be quite valuable as it enables both re-usability and scaling. For simplicity, a fixed set of initial conditions is used when evolving cellular automata rules.

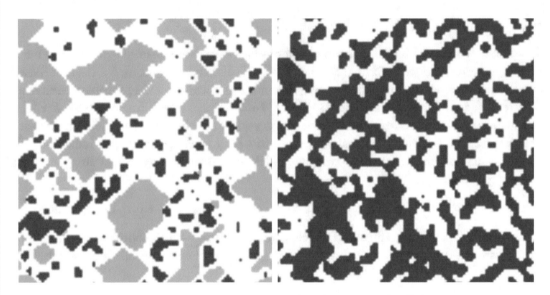

Figure 5.2: Examples of cavern-like level maps produced by evolved cellular automata.

The locality property creates a beneficial situation. The automata rules are evolved to yield a good map for a fixed initial set of conditions. These initial conditions are an assignment of randomly chosen cell states to the automata's first time step. The automata then updates the cell state several times, creating the map. It turns out that a good map, with a very similar appearance, is obtained for almost all sets of initial conditions; this is a remarkable robustness of the fashion-based cellular automata representation. The practical consequence of this is that a rule, evolved on a modest sized grid, may be used to created connected cavern maps on almost any size of grid, and that novel maps may be created by generating different initial state for the automata.

Cellular automata have been applied to the study of a diverse range of topics, such as structure formation [32], heat conduction [33], language recognition [54], traffic dynamics [48], modeling of biological phenomena [36], and cryptography [5], to name a few. Cellular automata have also been used for image and sound generation. Serquera and Miranda of the Interdisciplinary Center for Computer Music Research, UK, have many publications on the use of cellular automata for sound synthesis [1, 61]. Much of their work consists of mapping the histogram sequence of a cellular automata evolution onto a sound spectrogram, which produces spectral structures that unfolds in a patterned fashion over time. The authors claim that the mapping produces a "natural" behavior, and can replicate acoustic instruments [62].

Cellular automata have also been applied in the arts. They have been used to produce artistic images [25, 52], and their use has been extended to the fields of architecture and urban design [35, 63]. An interesting application has been the use of cellular automata in simulating the emergence of the complex architectural features found in ancient Indonesian structures, such as the Borobudur Temple [64]. Ashlock and Tsang [25] produced evolved art using one-dimensional cellular automata rules. These systems produced aesthetically pleasing images. In [21] a good deal of information about the fitness landscape of a particular type of cellular automata was derived.

This chapter introduces a novel type of cellular automata that is relatively easy to represent for evolution. The cellular automaton uses a toroidal two-dimensional grid as its cell space. The *von Neumann* neighborhood, of the sort shown in Figure 5.1, is used. The cellular automaton is synchronous with an updating rule governed by a score matrix that defines the benefit a cell has from having neighbors of each possible cell type.

At a given time step, the score each cell gets against its four neighbors is computed and summed, giving each cell a score. Each cell adopts the type of its neighbor, including itself, with the highest score. Because the cellular automaton is synchronous, each new cell state is computed, and then the state of all cells is updated simultaneously. When a cell is tied with a neighbor or neighbors for the highest score, the cell's state does not change.

Representation for Fashion-Based Cellular Automata

With $n = 6$ cell states, the rule for the cellular automaton is given by a 6×6 real score matrix that is specified by 36 real parameters. The matrix is encoded by a vector of 36 real numbers in the interval $[0, 2.0]$, specifying the entries of the matrix row-by-row. Since scoring is relative, the range of values is not critical. This representation groups the scores obtained by a single cell type together. Row i of the matrix consists of the score a cell in state i obtains against opponents of type $1, 2, \ldots, 6$.

The evolutionary algorithm used to search the space of score matrices employs two-point crossover, exchanging randomly chosen middle segments of the vector of real values. Point mutation is performed by modifying one of the numbers by adding a uniformly distributed random variable in the interval $[-\epsilon, \epsilon]$ with $\epsilon = 0.1$. If a value leaves the interval $[0, 2]$ as the result of mutation, a new value is generated uniformly at random in the range $[0, 2]$. The entries of the vector are the rows of the matrix, meaning that contiguous groups of six entries specify the scores one cell state obtains when matched against others.

It is important to note that there are many possible representations for cellular automata. This chapter uses a small group of closely related representations and does not explore the diversity of representation that is available. The representation used here is both quite simple and capable of expressing a large number of distinct behaviors.

Evaluation Tools

When using evolutionary computation as an optimizer, it is difficult to determine if the algorithm has been run for a sufficient time. In this study the evolutionary algorithm was permitted 20,000 fitness evaluations (2 per mating event). Eleven different sets of parameters were run for comparison. It would be interesting to have some idea which were closer to having converged. The *time of last innovation* (TLI) assessment can provide some perspective on this issue.

For all of the experiments we examine the fitness tracks of the experiments and compute the last time the current best fitness changes. This time of last innovation is then divided by the time the algorithm ran to obtain a *fractional time of last innovation*. Figures 5.3, 5.4, and 5.5 include results of the TLI evaluation. The TLI statistic serves as a much more compact way of displaying information about algorithm innovation behavior as opposed to the more traditional maximum-fitness-over-time plots, commonly used to assess if an algorithm was run long enough.

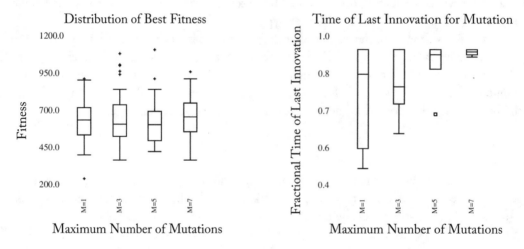

Figure 5.3: Shown are box plots of the distribution of best fitness values and the fractional times of last innovation for the mutation rate parameter study.

5.1.1   DESIGN OF EXPERIMENTS

The evolutionary algorithm used is steady state [67]. The model of evolution used is seven single tournament selection. In this model of evolution, seven members of the population are selected uniformly at random. The two most fit members of this group are copied over the two least fit and the copies are subjected to crossover and mutation. The size of seven is a compromise value between weak and strong selection. Mutation consists of a number of point mutations selected uniformly at random in the range 1 to M, the *maximum number of mutations*. The default value for the maximum number of mutations is $M = 3$.

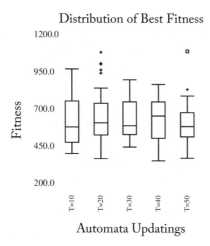

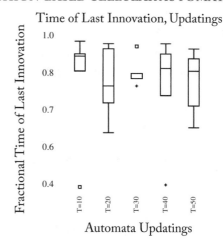

Figure 5.4: Shown are box plots of the distribution of best fitness values and the fractional times of last innovation for the number of automata updatings parameter study.

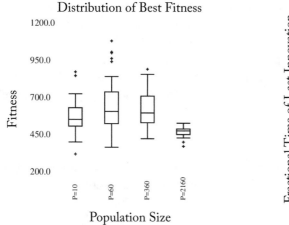

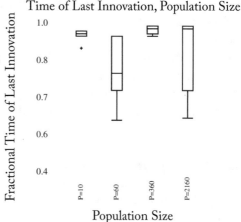

Figure 5.5: Shown are box plots of the distribution of best fitness values and the fractional times of last innovation for the population size parameter study.

The algorithm operates on a population of cellular automata rules with a default population size of $P = 60$. A single set of initial conditions on a 100×100 grid is filled in uniformly at random and then saved for used in all fitness evaluations. Cellular automata are evaluated for a default of $T = 20$ updatings of the cell states, and then fitness is assessed. A parameter study, varying the default values, was performed. The values used, varying one value at a time, are taken from $M = 1, 3, 5,$ and $7; T = 10, 20, 30, 40,$ and $50; P=10, 60, 360,$ and 2160.

The evolutionary algorithm is run for 10,000 instances of tournament selection, called *mating events*. Every 100 mating events summary fitness statistics, including mean, variance, and maximum are saved, as is the fraction of the population that has fitness zero. The definition of the fitness function includes an explanation of how these zero-fitness individuals arise.

The Fitness Function

After T updatings of the fashion-based cellular automaton rule from the fixed initial state the cell states of the cellular automaton are used to evaluate fitness. Before fitness is evaluated, a single application of a majority rule, like that used in [43], is applied to eliminate isolated pixels. This has little effect on the fitness but yields a better looking map. The state 0 is taken to represent empty space, all others are taken to represent obstructed cells. Using a recursive fill from a cell in the center of the state space, the number N of empty cells that can be reached from that central cell are computed. The fraction of unobstructed cell states U, accessible or not, are also computed. The fitness of an automaton rule is:

$$fit = \frac{N}{1 + |2 * U - 1|} \tag{5.1}$$

This function rewards cells accessible from the central cell of the grid and penalizes the grid if it does not have half its cells in state 0 (unobstructed). This encourages a half-unobstructed cell space much of which is connected. If the central cell used for the recursive fill is obstructed, then a rule receives a fitness of zero. Evolution rapidly eliminates such individuals, so this arbitrary choice of a starting point for the recursive fill is not a problem and leaves the evolution code simple. In use for level generation for a game, it would be worth finding the largest connected component, rather than starting at a fixed point.

Experiments Performed

A collection of 11 experiments, each consisting of 30 independent runs of the evolutionary algorithm, were performed. The parameter values used in these experiments are shown in Table 5.1. The experiments are based on the default parameter values $M = 3$, $T = 20$, and $P = 60$, chosen with preliminary experimentation during testing and debugging of the code. The experiments form three parameter studies. Experiments 1, 2, 3, and 4 compare different maximum mutation rates; Experiments 2, 5, 6, 7, and 8 compare different numbers of updatings of the fashion-based cellular automata during fitness evaluation; and Experiments 2, 9, 10, and 11 compare different population sizes. The default parameters are those used in Experiment 2 which is why it appears in all three parameter studies.

5.1.2   RESULTS AND DISCUSSION FOR CELLULAR AUTOMATA LEVEL CREATION

Examining Figure 5.3 it is not difficult to see that the maximum number of mutations is a relatively soft parameter. Juxtaposing the fitness information with the TLI data it is clear that

Table 5.1: Shown are the values for the experimental parameters used in experiments. The parameters varied are: maximum number of mutations M; automata updatings T; population size P.

Experiment	M	T	P
1	1	20	60
2	3	20	60
3	5	20	60
4	7	20	60
5	3	10	60
6	3	30	60
7	3	40	60
8	3	50	60
9	3	20	10
10	3	20	360
11	3	20	2160

the algorithm converges sooner when the maximum number of mutations is lower. This suggests that the algorithm is undergoing greater exploration when the number of mutations is higher (not surprising), but given the lack of impact on fitness that the exploration is not particularly useful. It probably consists of minor variations that make a few more cells accessible. The best balance of final results and convergence weakly supports the default choice of $M = 3$.

When examining the number of updatings of the cell states of the cellular automata during fitness evaluation, there is a strong motive to favor smaller values of T: time to evaluate fitness varies directly with T. Given that there is a lack of fitness variation between the different values of T, there is a temptation to set $T = 10$. The results of varying T, both fitness and time of last innovation, are shown in Figure 5.4. Examining the TLI data shows that $T = 30$ has anomalously early convergence—probably an actual anomaly and something that should be checked in the future. The results for $T = 10$ show the worst convergence behavior—suggesting that the randomness in the initial conditions is still strongly influencing the final cave-like level that evolves. This leave $T = 20$ as the best choice, balancing second lowest fitness evaluation cost with good convergence behavior. It is also worth noting the high-fitness outliers for $T = 20$.

The only parameter for which the fitness data returned significant differences is the population size study which provided clear evidence that a very large population is not a good idea. The best performance was turned in by the default value of $P = 60$ and, of the three populations sizes with very similar fitness results, had the best convergence behavior as measured by the TLI results. Overall the three parameter studies support the chosen default values $M = 3$, $T = 20$, and $P = 60$ in the weak sense that no better values are available from the perspective of fitness, and the chosen values exhibit superior convergence behavior.

Diversity of Results

The algorithm located a broad diversity of results. A sampling of the cave-like levels evolved are shown in Figure 5.6. Monochrome or almost monochrome levels are the most common and are more common for larger values of T. The last image in the second row is unusual in having horizontal and vertical corridor-like structure. All of the structures in the second row have a more block-like structure which is relatively rare. The last two rows of Figure 5.6 were chosen for their polychromatic character.

This remarkable diversity of results suggests that there is room to add additional constraints to the fitness function. This could include the presence of large open spaces, other levels of filled-cell density than the current 50%, and the use of check-point driven fitness like that in Chapter 2. The fitness function was chosen to reward having a large connected component and have the space in a level map 50% obstructed. This fitness function in no way encodes the idea of having a "cool looking" map. This means that, as with many evolutionary design systems that perform ACG, the local optima of the fitness landscape may be more desirable from the perspective of a game designer than global optima.

The Advantage of Rule Locality

In the experiments performed in this study, a single set of initial conditions was used for all fitness evaluations. This was done to permit repeatability, from a random number seed, of the experiments in case they needed to be rerun. The experiments used a 100×100 cell space meaning that one set of initial conditions is statistically similar to another, differing only in details. This, in turn, has an additional advantage. Once an automata rule has been located that creates a level with a desirable character, then any number of levels with a similar character may be generated by using new initial conditions. Examples of eight similar level maps for two different evolved automata rules are shown in Figure 5.7. In addition to being able to generate multiple maps from any given rule, the fact that the cellular automata act locally by neighborhoods means that we can change the size of the grids and get larger or smaller maps. Figure 5.8 shows an example of this.

5.1.3   DISCUSSION FOR CELLULAR AUTOMATA LEVEL DESIGN

The material presented thus far has demonstrated that fashion-based cellular automata can evolve a broad variety of cavern-like level maps. The parameter study yielded little practical advice, other than avoiding very large population sizes, and demonstrated that the system—at least when run with 10,000 mating events, is relatively robust to the choice of parameters. Taken as a whole, the TLI data suggests that, good results notwithstanding, running the system for a longer time might yield better fitness values.

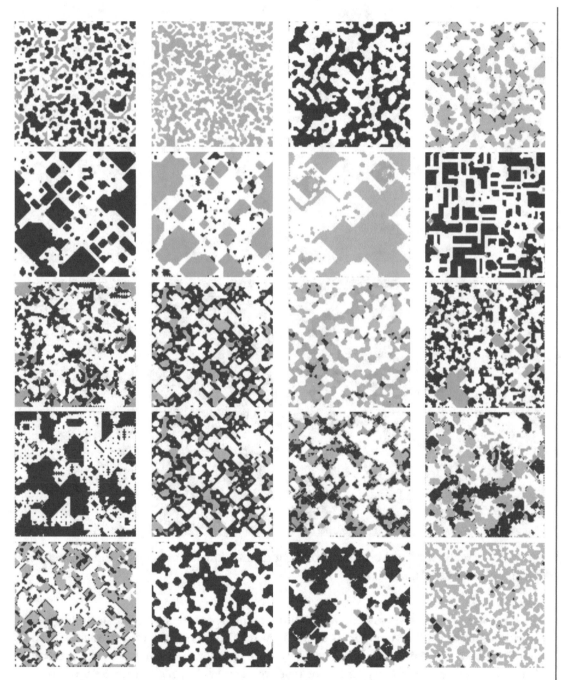

Figure 5.6: Examples of final cell states rendered as level maps.

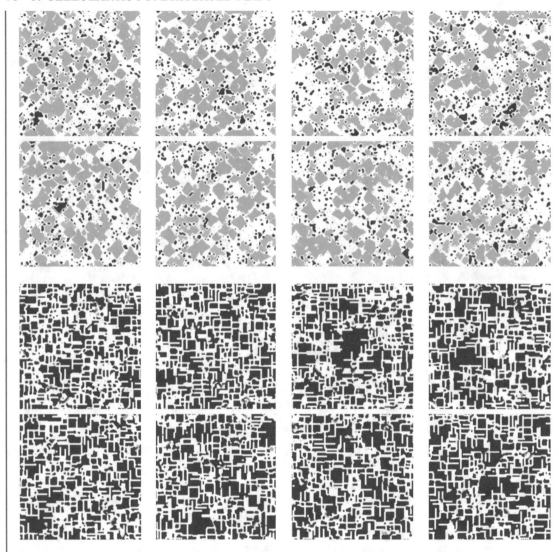

Figure 5.7: Examples of final cell states for eight distinct initial conditions using two examples of fashion-based cellular automata.

The use of six states, only one representing empty space, was a choice made to give the algorithm greater expressive power. If a level map requiring only one type of rock is needed, then the solid states can simply be merged into a single color. In many cases the obstructions of different color could represent different types of obstructions: rock, water features, pits, or even pools of lava. Fully exploiting this potentiality would require writing more complex fitness functions.

Figure 5.8: A 400 × 400 version of the second type of automata rule shown in Figure 5.7.

An early priority for additional research is to investigate the impact of changing the number of cell states. This study did not have room to encompass this variation of the system and so six were chosen as being a relatively high number that enables expressibility. Two or more cell states would permit the system to function; larger numbers of cell states enable enormously larger spaces of rules with correspondingly larger sets of potential appearances.

5.1.4 USING AN OPTIMIZER FOR NON-OPTIMIZATION GOALS

The evolutionary algorithm used in this section is structured as an optimizer. It is optimizing a simple fitness function that encodes some desirable properties of a level map. The fact that almost all runs of the algorithm produced a cavern-like level, however, follows from the choice of the fashion-based cellular automata representation. In particular, the property that this representation preserves homogeneous regions encourages caverns. Since we are searching for levels that would please a human designer, we *are not optimizing* with our optimizer; we are directing search to perform procedural content generation.

In 3,300 runs of the evolutionary algorithm no duplicate maps appeared. This was checked by performing pixel-level comparison for a fixed set of initial conditions. This means the fitness landscape is exceedingly rich in local optima. From the point of view of providing a tool to a human designer, this is a good thing. It also means that, as an optimization problem, this fitness landscape is challenging. Finding a global optima would be exceedingly difficult and not too useful.

5.2   GENERALIZING FITNESS AND MORPHING

In the previous section we created a single fitness function that generated cavern-like maps with about half the space empty and a large connected component. In this section we will examine a variation on the fitness function and a new way to use the score-matrix representation for fashion-based cellular automata.

5.2.1   GENERALIZING THE FITNESS FUNCTION TO CONTROL OPEN SPACE

The fitness function used in the previous section to judge the quality of an automata rule M is:

$$f(M) = \frac{Z}{1 + |1 - 2U|}, \tag{5.2}$$

where Z is the number of cells in state zero in the connected empty component of the map including the middle cell of the map, and U is the fraction of empty cells. If the center cell is not in state zero, this function yields a fitness of zero. This function rewards a large connected network of open spaces and a map that is 50% full. We will briefly examine the impact of changing this function to

$$f(M) = \frac{Z}{1 + |1 - \frac{U}{\alpha}|}, \tag{5.3}$$

where α is the desired fraction of full states, to change the density of the resulting map. Note that low values of α make it easy to have a large connected component of empty space, and so yield less challenging fitness functions. Using the software from [8] with different values of alpha, it is possible to obtain maps like those shown in Figure 5.9. These are minor new results using small modifications of the original fitness function.

5.2.2   RETURN OF DYNAMIC PROGRAMMING BASED FITNESS

Next we will apply a fitness function, not used before on cellular-automata based level maps, but shown to be useful for decomposing the level map design problem [15]. This fitness function requires that openings to the cavern system be present in a manner that permits the evolution of *tiles* that can be linked to form larger maps. The tile-linking techniques are demonstrated in Chapter 5. For this study an opening in the middle of each side of the map is required, and it is

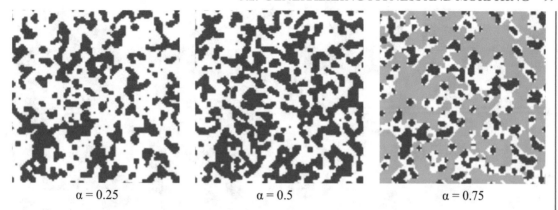

$\alpha = 0.25$ $\qquad\qquad\qquad$ $\alpha = 0.5$ $\qquad\qquad\qquad$ $\alpha = 0.75$

Figure 5.9: From left to right these maps were generated by rules that had target densities of full cells of $\alpha = 0.25, 0.5,$ and 0.75.

required that there be an open passage joining each pair of openings. If either of these constraints are violated, the automata rule receives a fiat fitness of zero. Otherwise the fitness of the rule is the average of the six distances between pairs of openings, computed with simple dynamic programming. An example of the types of tiles that evolve is shown in Figure 5.10.

A Parameter Study of Dynamic Programming Fitness

A total of 161 experiments were performed in which various parameters were modified to check their impact on fitness. These parameters were tournament size, number of mating events, population size, and the maximum number of mutations used. The number of mutations applied to a new structure is selected uniformly at random in the range one to MNM, the *maximum number of mutations* parameter. The population size was changed in two batches. For the first batch of population size experiments, population size was set to 60, 360, 370, 400, 500, 1,000, 2,000, 3,000, 4,000, 5,000, and 10,000. The irregular spacing of population sizes was the result of trial-and-error tuning of the population size. The two very close values, 360 and 370 appeared very different in initial runs, but not once the full data set had been collected. Then for each of these values the MNM value was changed from 1 to 12, resulting in 120 experiments being run. For the second batch of population size experiments, population size was assigned values of 50, 100, 200, and 300. Then for each of these values the MNM value was set to 1, 4, 7, 10, and 12 resulting in 20 experiments being run.

For the next batch of experiments tournament size and mating events were changed. For tournament size the values tested were 5, 7, 8, 10, 15, 20, 25, and 30 resulting in 8 experiments being run. For the mating events the values that were tested were 5,000, 10,000, 11,000, 12,000, 13,000, 14,000, 15,000, 20,000, 30,000, 40,000, 50,000, 100,000, and 1,000,000 resulting in 13 experiments being run. A partial visualization of these results is given in Figure 5.11.

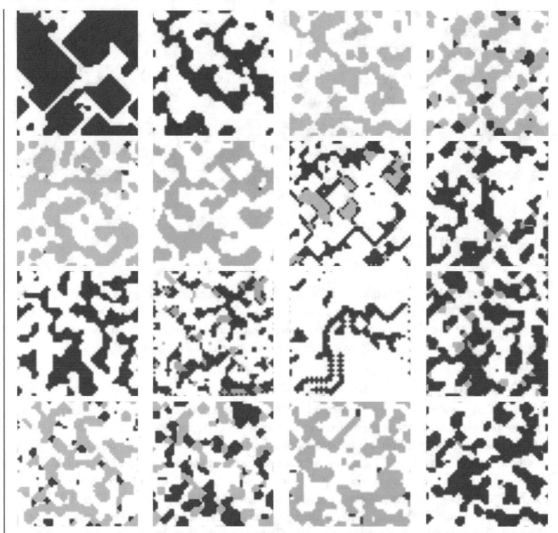

Figure 5.10: A collection of evolved cellular automata based tiles. Each has an an opening in the center of each tile face; these openings are mutually accessible via empty space.

The parameter study shows that a small population size and a low mutation rate are both favored. The provision of additional mating events, extending the time of evolution, had a small beneficial effect that seems to top out at 30,000 mating events. These results will inform future experiments.

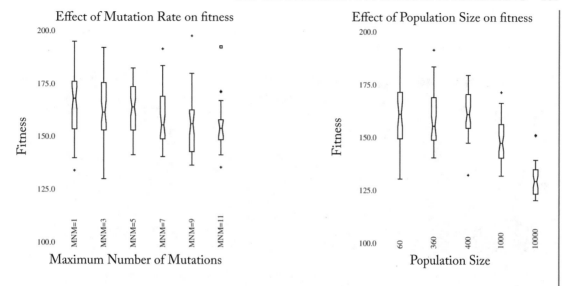

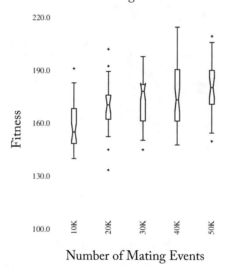

Figure 5.11: Shown are inflected box plots for the impact of mutation rate (top left), population size (top right), and duration of evolution (bottom) on the final fitness of the population.

Lack of Robustness in Dynamic Programming

The simplest fitness function for level map evolution, used in Chapter 1 for chess and chromatic mazes, was to make the maze as long as possible. Unlike the fitness function that requires that the cellular automaton generate a large empty space and otherwise fill space to some degree, the goal of a long path is not robust to change of initial conditions. Figure 5.12 shows a dynamic programming driven evolved automaton which maximizes the path from the upper left corner to the lower right.

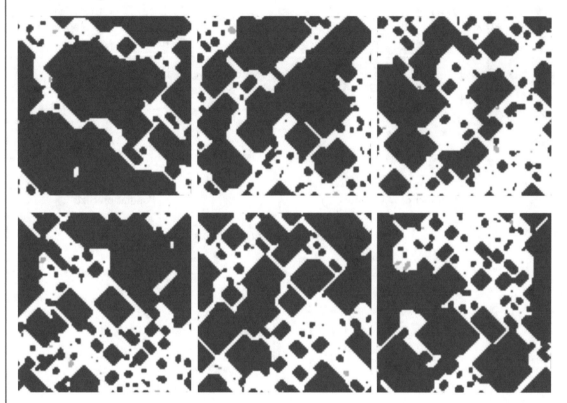

Figure 5.12: The automaton was evolved to create a path from the upper left corner to the lower-right of maximum length. The first panel is the automaton as it plays out with the fixed initial conditions. The next five panels are the same automata for distinct initial conditions.

The first panel in Figure 5.12 is the instance of the automaton that was used in fitness evaluation. This is a good map with a branching path. The other five panels are re-runs of the automaton rule on different initial conditions. The five re-runs contain only one automata that has open space at the upper left and lower right corners. Interestingly this map would have given the automata *higher* fitness than the one in the first panel. The maps in the other four panels earn fitness zero for failure to connect the upper left and lower right corners.

The resampling experiment demonstrates that, unlike the original cavern maps, the dynamic programming based automata have a fitness that is commandingly dependent on the initial conditions, losing much of the robustness that the original representation possessed.

5.2.3   MORPHING BETWEEN RULES

The local nature of automata updating, quite useful when replayability and scalability are issues, nevertheless leads to a problem with relative uniformity of appearance. If you make a large map with a particular automata rule, then different pieces of it will look quite similar to one another in their use of states (colors in the renderings shown here), connectivity, and shapes of the solid spaces that appear. *Morphing* provides a technique for avoiding the relative uniformity of appearance.

A representation for use in evolutionary computation is *morphable* if convex combinations (weighted averages) of instances of the representation are instances of the representation. The real-valued matrices used to specify fashion-based cellular automata clearly have this property.

A *morph* from the rule specified by matrix M_1 to the rule specified by matrix M_2 is the line segment in rule space given by

$$(1 - \lambda)M_1 + \lambda M_2 \ \ 0 \leq \lambda \leq 1$$

A morph is a linear space of rules, and our second conjectural application, re-evolvability, will attempt to exploit this in a future study. Suppose, however, that we permit the rule used to update the automata to vary spatially, deriving the value of λ to vary based on the position within the cell matrix. Then we can obtain results like those shown in Figure 5.13.

The maps in Figure 5.13 were selected by examining the 435 possible pairs of morphs between distinct best-of-run (most fit) rules from 30 runs of the baseline algorithm. Two sorts of morphs were performed, *radial* and *lateral*. The upper panels of the figure are radial morphs, where λ is zero at the center of the picture and one in the corners; the lower panels are lateral morphs where λ is the fraction of the distance across the picture in the horizontal direction.

Some of these resulting maps are empty in the regions where λ is not near one or near zero and many of the maps exhibit a solid, impassible zone even more severe than the one in the bottom left panel of Figure 5.13. The left radial morph is subtle—features appear near the center of the picture that do not appear elsewhere, while the right radial morph shows a large change in the character of the caverns from the middle to the edge of the diagram.

Enhancing Morphability

The advantage of morphing between a pair of rules is that it opens up a wider range of possible appearances for level maps, potentially solving the issue of lack of long-range variability in appearance. The down side of morphing, at least thus far, is that many of the morphs between individually evolved rules are not usable maps; they would receive a fitness of zero. There are at least two possible ways to address this problem.

The first method for avoiding unusable morphs is to evolve pairs of rules that are directly tested, during fitness evaluation, for their ability to generate useful morphs. There is a potential danger here: morphing from a cellular automaton to itself. This would generate a morphable pair of rules with a single local appearance. Different matrices can specify the same rules and so,

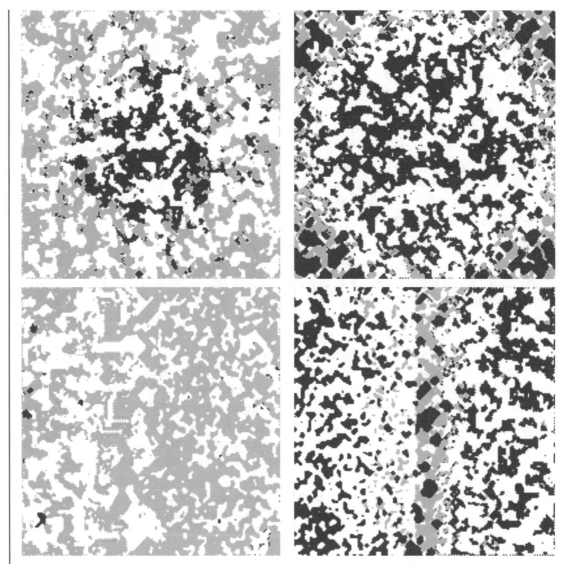

Figure 5.13: The top two maps permit the morphing parameter to operate radially, the bottom two laterally.

in addition to simultaneously evolving two automata rules, the fitness evaluation must include provisions to avoid the large local optima that involves evolving two, essentially duplicate, rules. The presence of multiple states representing filled in cells provides an easy way to do this—simply require that the density of types of filled states be distinct in areas largely governed by each of the automata rules.

The second method of finding morphs involves the use of *fertility* testing. Here we generate a large collection of cellular automata and test their pairs for viable morphs. The cellular automata are then ranked by the number of other automata rules with which they produced viable morphs. The highest scoring cellular automata in this sense have an enhanced probability of producing viable morphs. These could be used directly or might be an excellent starting population for the evolution of morphable cellular automata.

5.2.4 MORE GENERAL APPLICATION OF MORPHING: RE-EVOLUTION

We now have the notion of convex representations, those where the average of instances of the representation are still instances of the representation. These are exemplified by the cellular automata rules used in this chapter, but there are many others. They include any real-parameter optimization problem from standard function optimization to location of the weights of a neural net.

Suppose that we have a convex representation. Then for a set of α_i with the property that $\sum_i \alpha_i = 1$, we have that

$$\sum_i \alpha_i R_i$$

where R_i is an instance of a morphable representation. The set of possible weights α_i are *themselves* a representation of a subspace of the representation from which the R_i are drawn. In effect, if a representation is convex, we may transform the search for members of that representation into a real-parameter optimization problem where the real parameters are the averaging weights α_i.

The selection of the R_i represents a huge opportunity. While this is not an exhaustive list, these possibilities open up when this sort of *re-evolution* is used.

- If the number of R_i used is smaller than the dimension of the original morphable representation, then re-evolution is a functional method of dimension reduction.

- Considerations of dimension aside, the use of re-evolution is a *re-parametrization* of the problem with the R_i serving in the role of examples that anchor the re-parametrization.

- If the number of R_i used is larger than the dimension of the original problem, then the space becomes many-one (if it wasn't already). Many-one spaces are often easier to search—they have a more tractable fitness landscape, rich in paths out of states that might be local optima in the lower-dimensional version of the search.

- In the event that the problem being solved is constrained in some way, using a set of R_i that satisfy the constraints may help enormously—this depends, however, on the degree to which the constraints are themselves convex.

The utility of re-evolution will clearly have substantial dependence on the exact problem to which it is applied, and it is also not usable on non-convex representations. Given these limits, it has substantial potential as a source of representations.

CHAPTER 6

Decomposition, Tiling, and Assembly

This chapter will leverage material from Chapter 2 to design large maps with user-specified features in an incredible hurry. The assembly process is detailed in Section 6.1; an original version of this work was presented in [18]. In Section 6.2 we show how to incorporate user-specified details into first tiles then assembled maps.

6.1 MORE MAPS THAN YOU COULD EVER USE

Assembly of tiles borrows from the author's earlier research [16] on the creation of dual mazes which were shown in Chapter 3. It is worth noting that, while the simple binary representation is used in this chapter, the full generality of strategies for controlling the character of mazes generated in the earlier chapters are available for controlling the assembly plans and generating tiles presented in this chapter. Let's begin by looking at an example of our goal. Figure 6.1 shows a map built from 36 tiles.

The dynamic programming algorithm that is the core of the fitness functions used to create tiles has a time complexity that scales as the number of grids in a map. Quadratic time is acceptable for generating tiles, but it is problematic for generating really large maps, particularly on devices that may not have sufficient computational power or if a large number of different maps are required to enable replayability. This makes decomposing maps into tiles a natural choice. In [18], small evolved height maps of the sort presented in Chapter 2 were used as assembly plans. For each grid in the height map, the connectivity between tiles was checked and tiles with appropriate openings were assembled. The connectivity of the height map became the global connectivity of the overall map.

The tiled map in Figure 6.1 was constructed with *full* connectivity. Each tile has a passage to its adjacent tiles. Once an assembly plan that controls the strategic character of a map is chosen with the position of openings to adjacent tiles for each tile, there must be a library of available tiles to permit the user to fill in appropriate tiles. Figure 6.2 shows the possible types of tiles—with the caveat that the six tiles shown are representatives of rotation classes.

A second feature that can be incorporated into the tiles is their character or density. If we place a checkpoint, as in Chapter 2, into each of the entrances, then we can optimize a tile for

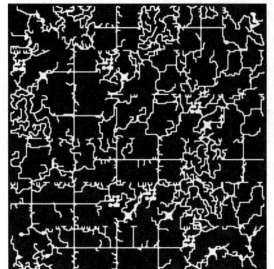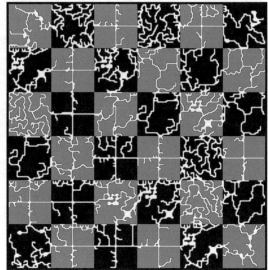

Figure 6.1: A 6 × 6 map, assembled from tiles, shown as a plain map and with the tiles marked.

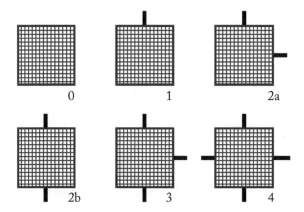

Figure 6.2: Shown are the six possible sorts of tiles. Type 0 tiles have no entrance. There are four possible rotations of Types 1, 2a, and 3. Tile Type 2b has two rotational types, while Type 4 tiles have one.

direct connection or long, winding paths. Figure 6.3 shows examples of tiles of Type 2b drawn from two sets of tiles with very different characters.

Let's look at a map with an assembly plan that requires a long winding top-level map. Figure 6.4 shows four versions of the map and the assembly plan. The tiles are required to have entrances based on their type, but they sometimes develop additional connections by chance.

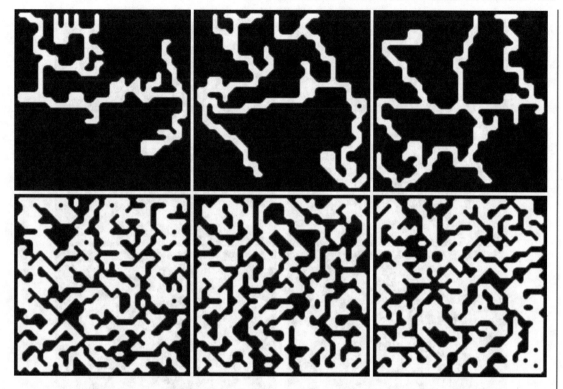

Figure 6.3: Show are six Type 2b tiles that are guaranteed to connect the left and right faces of the tiles.

These accidental connections can easily be eliminated by writing a slightly more stringent fitness function for tile evolution.

The tiles used in Figure 6.4 are sparse—they have relatively little open space. If we modify the code to place denser tiles in the center of the map then we get a figure like the one in Figure 6.5. This demonstrates how easy it is to leverage the different styles of tile created in Chapter 2 to permit assembly of large maps with planned characters.

The effective use of the plan-and-tile assembly strategy requires that the tactical properties of individual tiles be controlled carefully.

6.1.1   DETAILS OF TILE PRODUCTION

The fitness function used to create the tiles used in this section, the sum of distances between entrances as computed with dynamic programming, is considerably easier to understand than those used in Chapter 2. This algorithm can be used to generate libraries of tiles offline in an efficient way, seconding strategic considerations to the assembly plans. Since assembly plans are

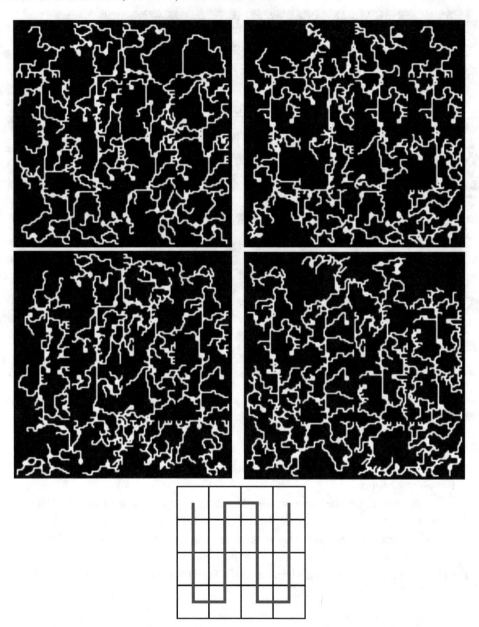

Figure 6.4: The blue line shows the required connections forming an assembly plan for a 4 × 4 map. Six instances of the map are shown. The blue line gives *required* connections, but other connections are possible by chance. Along the path the tile types are 1, 2b, 2b, 2a, 2a, 2b, 2b, 2a, 2a, 2b, 2b, 2a, 2a, 2b, 2b, 1.

Figure 6.5: This figure uses the same winding assembly pattern as the tiles in Figure 6.4 but selects from a library of denser tiles in the center of the map.

fairly small dual-height mazes they can be generated rapidly. These two tasks decompose the problem of level assembly for large levels in two offline tasks. Assembly itself is an algorithm with time linear in area of the final assembled level, making real-time assembly of level maps from pre-generated tile libraries an easy task.

The evolutionary algorithms used to build tiles use tournament selection of size seven [6]. Two-point crossover and uniform mutation with probability $p_m = 0.01$ were used. Mutated loci simply flipped the full/empty bit. For mutation of a height while evolving assembly plans, a Gaussian random variable with a mean of 0.0 and standard deviation of 0.5 was added to mutate a loci. Each tile-generating experiment was run 30 times and the most fit tile saved to obtain libraries of 30 tiles of each type.

6.1.2   ENUMERATING MAPS AND EXPLOITING TILE SYMMETRIES

Suppose we are assembling tiles into a 4×4 map of the sort shown in Figure 6.5. Each of the 16 locations has a tile of a particular type, and for each type the evolutionary algorithm has 30 instances of tiles of that type. This gives us

$$30^{16} = 430,467,210,000,000,000,000,000$$

possible maps. As we've seen, however, tiles of a particular type can have multiple styles. In fact each tile type has another 30 versions (there are dense and sparse tiles of each type) so we actually have

$$28, 211, 099, 074, 560, 000, 000, 000, 000, 000$$

maps. If, on the other hand, we wanted to tile an 8 × 8 map there would be 633, 402, 866, 629, 732, 777, 061, 622, 869, 468, 118, 866, 098, 964, 618, 280, 960, 000, 000, 000, 000, 000, 000, 000, 000, 000, 000, 000, 000, 000, 000, 000, 000, 000, 000, 000 or 6.3×10^{114} maps available for quick assembly. You get the idea: more maps than you could ever use.

Oddly, there is an easy way to get *even more* tiles that will pay a benefit in the next section. Square tiles have eight symmetries (including the identity in which nothing moves) that are given by the *dihedral group of the square* [40]. Figure 6.6 shows the eight symmetries of the square. The square labeled "start" is the initial state of a given tile. The first row is made of of the four rotations (including "don't"). The second row consists of the four ways to flip the tile over. These symmetries on the square act differently on different types of tiles.

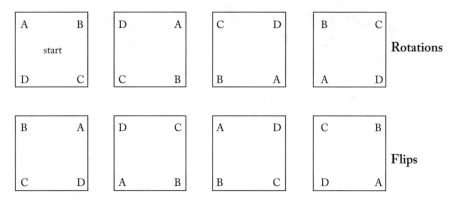

Figure 6.6: Shown are the four dihedral symmetries of the square.

For squares of Type 2a, the four rotations are needed to position the entrances properly. This means, however, that the eight available symmetries yield two versions of each entrance placement. If the entrances start in the correct place, for example, the "don't" symmetry that moves nothing *and* the flip that exchanges the positions of the two entrances both yield correct placements of the entrances. The number of different symmetric partners of a tile are two for tiles of Type 1, 2a, and 3. There are four possible symmetric partners for tiles of Type 2b and eight for tiles of Type 4.

6.2   REQUIRED CONTENT

Thus far the tiles produced for assembly have had constraints on where entrances must be, the rough density of filled squares, and on the distances between entrances. All of these can be managed well by dynamic programming. In this section we augment the technology to permit the inclusion of designer content into the tiles. This work is adapted from [50].

We retain the 33 × 33 grid used for tiles in the first section. This grid size is chosen because it permits the symmetric placement of a door in the middle of one side of a tile or two doors, at positions 10 and 21, equally spaced along the side, and is large enough to permit interesting structure without, itself, being unmanageably large. Grids with twin entrances on one side are not used here but appear in [18]. A feature specification for a type of tile gives:

1. the number and position of entrances to the tile;

2. the number of required features;

3. a shape for each required feature, including the position of checkpoints; and

4. a specification, for each pair of checkpoints, if the distance between them is to be (i) minimized, (ii) left free, or (iii) maximized.

Figure 6.8 shows a shape specification for a required feature, a square room with four pillars. This feature appears four times in the map in Figure 6.7. Required features must be rectangular and filled with four symbols. The symbols have the following meaning: 0 – empty space, 1 – filled space, 2 – space to be filled in by an evolvable bit later in the representation, 3 – empty space that is a checkpoint. This latter symbol permits the designer to place non-rectangular shapes within a bounding rectangle.

The bit string that specifies if grids within a tile are open or full starts with a sequence of integers in the range $0 \leq n \leq 10{,}000$. One such integer is included for each required feature. These integers n are decoded to positions within the 33 × 33 tile using

$$x = n \bmod (33 - D_x) \tag{6.1}$$

$$y = (n/(33 - D_x)) \bmod (33 - D_y), \tag{6.2}$$

where D_x and D_y are the dimensions of the required feature. The required feature is then copied into the tile at that location. There are enough bits, after the required feature position loci, to cover the entire grid. Any location which is already open or filled because of a required feature ignores the bit; otherwise the bit determines if the square will be open or filled. In the event that two required features overlap, or include the grid square containing an entrance, a tile is assigned a fitness of zero.

6.2.1   THE FITNESS FUNCTION FOR REQUIRED CONTENT TILES

A total of 24 experiments were performed to create tiles. The first 20 consisted of 5 groups of four for each of the 5 tile patterns with entrances given in Figure 6.2.

- The first two in each group of four had required content consisting of single checkpoints or 3 × 3 rooms which have little visual impact on the tiles.

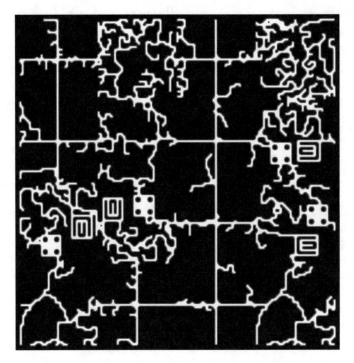

Figure 6.7: An example of a 4 × 4 tiled map with required content.

```
9 10
2 2 2 2 2 2 2 2 2
0 0 0 0 0 0 0 0 0
0 1 1 0 0 0 1 1 0
0 1 1 0 0 0 1 1 0
0 0 0 0 0 0 0 0 0
0 0 0 0 3 0 0 0 0
0 0 0 0 0 0 0 0 0
0 1 1 0 0 0 1 1 0
0 1 1 0 0 0 1 1 0
0 0 0 0 0 0 0 0 0
```

Figure 6.8: Shape specification of a 9 × 10 required feature. A square 9 × 9 room with a checkpoint in its middle and a north wall that is left free for the bitwise portion of the algorithm to fill or leave open.

- The second two of each group incorporated one or more rooms with a complex shape that has a substantial visual impact.

The last four experiments all generated tiles with four entrances.

- The first two were coerced to yield tiles symmetric about both the horizontal and vertical central line of grids in the tile.

- The last two were coerced to yield tiles with four-fold rotational symmetry.

In both cases this was accomplished by reducing the length of the gene to cover one-fourth of the tile and filling in the rest of it with the imposed symmetry.

A total of six fitness functions were used. M is the sum of distances between checkpoints whose mutual distance is to be maximized; m is the sum of distances between checkpoints whose mutual distance is to be minimized; C is the number of culs-de-sac; and F is the number of filled grid squares.

The fitness functions are:

$$F1 = \frac{M+1}{m+1} \tag{6.3}$$

$$F2 = \frac{M+1}{m+1} + C/5 \tag{6.4}$$

$$F3 = \frac{M+1}{m+1} + C/4 + F/20 \tag{6.5}$$

$$F4 = \frac{M+1}{m+1} * \arctan(C) \tag{6.6}$$

$$F5 = \frac{M+1}{m+1} * \arctan(F) \tag{6.7}$$

$$F6 = \frac{M+1}{m+1} * \arctan(F) + C/5 \tag{6.8}$$

$\arctan(x)$ is a squashing function that reduces the marginal reward for more filled squares or culs-de-sac. These functions were chosen by running several preliminary experiments and choosing the squashed or linear reward that looked best. The details of the experiments are specified in Table 6.1.

A key point is that the pattern of requested maximized, minimized, and free distances between checkpoints gives the designer substantial tactical control over the properties of the

Table 6.1: Specification of experimental details. Type is the tile type; C is the number of checkpoints; FF is the fitness function; Xp is the number of pairs of checkpoints for which distance is maximized; Np is the number of pairs of checkpoints for which distance is minimized.

Exp#	Type	C	FF	Xp	Np	Exp#	Type	C	FF	Xp	Np
1	1	4	6.5	3	4	13	3	4	6.5	3	3
2	1	4	6.5	3	5	14	3	4	6.5	4	0
3	1	5	6.5	3	4	15	3	4	6.5	3	3
4	1	5	6.5	3	4	16	3	4	6.5	3	3
5	2b	6	6.5	7	2	17	4	5	6.5	8	2
6	2b	6	6.3	4	2	18	4	10	6.5	12	3
7	2b	4	6.8	1	3	19	4	6	6.8	2	8
8	2b	4	6.8	4	3	20	4	6	6.8	2	8
9	2a	6	6.5	7	2	21	4	4	6.4	4	2
10	2a	6	6.5	3	6	22	4	4	6.6	0	6
11	2a	4	6.8	2	3	23	4	4	6.6	2	4
12	2a	4	6.8	2	3	24	4	4	6.7	4	4

resulting tiles. The control is sufficient to ensure that any tile in a library evolved for a given specification can be swapped in for another without more than minor changes in the connectivity and relative distances within the map built from the tiles. As we have seen, given an assembly plan with desired properties, it is therefore possible to rapidly assemble vast numbers of maps that realize the assembly.

Suppose that a game designer has particular encounters: a boss encounter, a crypt full of ghouls, the tomb of a vampire, a dragon's treasure horde. These can be included in a tile design as required features. The placement of the encounter within the map can be held constant, or localized to a small area of the assembly plan, but the details of the tile containing the encounter can be varied by generating a large tile library for the specification containing the feature.

Choosing to maximize distances between some pairs of checkpoints and minimize distances between others, when the paths that realize those distances live in the same two-dimensional tile, creates conflicting goals that yield a fitness function rich in local optima. The final fitness of the best tile varies substantially within each experiment, suggesting that most of the runs are ending in local optima of the fitness function. The fitness functions themselves were chosen based on intuition about what elements were important to producing good tiles combined with trial and error to find tiles that, in the end, looked good to the researchers.

6.2.2   RESULTS OF THE TILE CREATION EXPERIMENTS

An example tile from each of the 24 experiments is shown in Figure 6.9. Note the substantial variety of tile types and densities that can be obtained with a fitness function utilizing only three types of measurements of a tile (inter-checkpoint distance, number of culs-de-sac, number of filled grid squares). Rewarding culs-de-sac causes there to be more dead ends in the tile. Rewarding filled grid squares increases the density of the tiles. Requesting that distances between checkpoints be maximized yields winding connecting paths, while asking that they be minimized favors direct paths. Since checkpoints in required features can appear in various locations, the rough location of a feature can be controlled by asking that its checkpoint be far from one entrance but near another. The relative positions of required features representing rooms is also, to some degree, controlled by requests for long distances or short distances between those checkpoints.

In evolutionary computation the goal is typically to find global optima while escaping local optima. In [12] it is noted that, when using an evolutionary algorithm to locate aesthetic images, local optima can be helpful. A fitness function that attempts to estimate the aesthetic quality of an image is, in all probability, an approximate guess at a measurement of the quality actually desired. This means that a human judge may rate what the fitness function thinks are locally optimal images above globally optimal images. The fitness functions used to evolve tiles in this study have a similar character to those used in evolved art. While the fitness function grants control of tactical detail, some of the results look better than the others in ways that are difficult to accurately quantify. This means that an algorithm that is not too zealous in the pursuit of global optima simply enriches the tile set substantially.

We conclude the figures for this section with six examples, shown in Figure 6.10, of different 4 × 4 maps assembled from different tile sets with different assembly plans—tiles without symmetry, rotationally symmetric tiles, dense and sparse tiles, and tiles with required content. The material in this section is a starter set on the possibilities for tile design and assembly. Fitness constraints, possible features used to define new fitness functions, and the entire issue of tactical assembly of the tiles still require substantial additional work.

The next test for the level generation technique should be to incorporate the generated levels into a game of some kind. There seem to be three natural venues. The first is an encapsulated, portable minigame where the tile itself, augmented perhaps with puzzles, would form a self-contained unit. The ability to have a minigame that has a different map each time it is activated would improve replayability. In this case puzzles could be drawn from work like that in Chapter 1 where barriers, rather than being explicit walls, are implicit in the behavior of traps or environmental features.

The second possible venue is to generate levels for a first person shooter or roguelike adventure game. This is a challenging enterprise requiring either substantial development or integration of the tile library and assembly code into existing platforms. The third is to incorporate

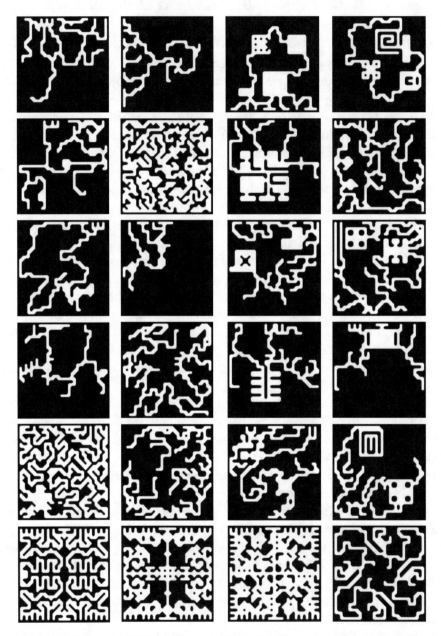

Figure 6.9: Example tiles from Experiments 1-24 in reading order. The tiles are bordered with black but, when used to assemble larger maps, a tile appears in the middle of each side that has an entrance.

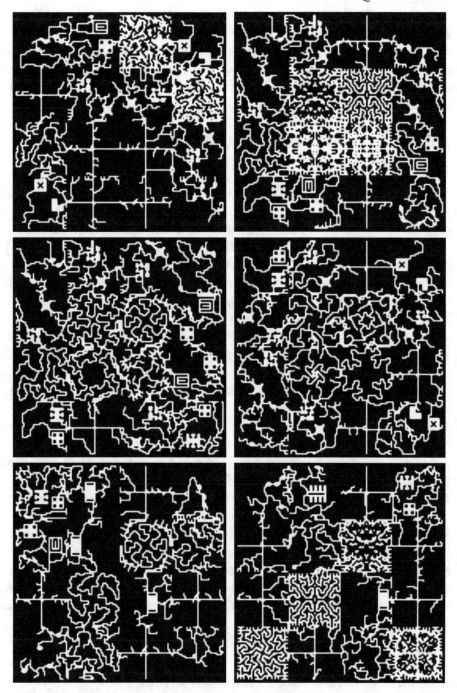

Figure 6.10: Examples of assembled tile maps using different tile sets and assembly plans.

the map generation technology into software that automatically generates modules for a fantasy role-playing game, the subject of the next section.

Anything from levels of a traditional fantasy role-playing dungeon to an endless supply of warehouses and industrial buildings for a zombie-apocalypse scenario are well within the map generation capabilities of the techniques given here. These techniques could also incorporate the sort of structures created in Chapters 3 and 4.

6.3 CREATING AN INTEGRATED ADVENTURE: GOBLINS ATTACK THE VILLAGE

This section presents software for generating what amounts to a single, quite complex tile. This tile is the complete map for a traditional starting adventure for fantasy role-playing games (FRPGs). In this adventure, a troop of goblins has attacked an unprotected village and the characters in the game, run by the people playing the game, are supposed to chastise the goblins. Unknown to the player, the goblins themselves have been intimidated into doing the will of an evil mage. The goblin's lair is in the same underground cave system as the mage's new digs, and these two things are persistent parts of the map being designed.

FRPGs have been commercially available for about 40 years with the initial offering of *Dungeons and Dragons* (D&D) (tm) appearing in 1974. D&D has become a generic term for role-playing games [44], much like Kleenex for tissue paper, but the field is far broader than the heroic fantasy genre that D&D simulates. Games exist that permit play in science fiction, pulp fiction, post-nuclear holocaust, superhero, cartoon, historical situations, and last-days apocalypse fiction.

Role-playing games also exist for popular television series and movies like *Buffy the Vampire Slayer* and *The Lord of the Rings*. Characters may be wizards and warriors, demons or angels, hard-boiled private detectives, superheroes, the slayer or her companions, or owner-operators of a Free Trader starship. Different systems and genres have strong partisans and the field has diverse content and appeal. Many gaming groups use commercially written adventures and there is a substantial market potential for automatically generated content.

This technique is highly scalable, but it produces a map with a constructively sparse type of connectivity in which there is one and only one path between any pair of rooms.

This section provides proof-of-concept for an automatically generated FRPG module with a number of points where the designer can exert control over the type of adventure module generated. Among the various sub-tasks required, the one that has been most studied is automatic map design.

6.3.1 SYSTEM DESIGN FOR FRPG MODULE CREATION

Automatically creating an FRPG module has a number of steps.

1. Select frame and required content.

2. Evolve level maps.

3. Identify rooms.

4. Extract room adjacency graph.

5. Populate rooms with monsters, treasures, and traps.

An example of a *required content* specification appears in Figure 6.11. This example required content is, in fact, the template for the rooms where the evil mage hangs out. A configuration file specifies items of required content which are placed in the level by an evolutionary algorithm. The *frame*, shown in Figure 6.12, is similar to a required content specification, but it is the same size as the level. The frame differs from the required content in that it places content that appears in the same position in all instances of evolved levels. In this study the frame is used to create an entrance area, seen as room 1 and the crossed corridors to the right of room one in Figure 6.13. The example of required content given in Figure 6.11, the evil mage's lair, appears as rooms 14 and 18 in Figure 6.13.

```
14 6 //size
11111111111111
10000100000001
10000000000002
10000100000001
00000100000031
11111111111111
```

Figure 6.11: An example of required content. Four sorts of cells are specified, 0 denotes empty space, 1 denotes filled space, 2 denotes a space that may be full or empty, as determined later by the evolutionary algorithm, and 3 denotes empty space that is recorded as a check point. The 2 in this example represents an optional entrance that evolution may realize as space or a wall.

6.3.2 THE LEVEL EVOLVER

In the example dungeon used in this study, a simple frame is used. It contains a single checkpoint denoting the entrance, realized as room 1 in Figure 6.13. The example dungeon also uses four pieces of required content. The dungeon is a small one, intended to be a single-level goblin garrison in a buried, abandoned building. In the example map the required content is realized as rooms:

1. Rooms 8, 9, and 10, form the goblin's lair,

2. Room 7, an armory from the days before the building was buried,

```
11111111111111111111111111111111111111111111111111
12222222222222222222222222222222222222222222222221
12222222222222222222222222222222222222222222222221
12222222222222222222222222222222222222222222222221
12222222222222222222222222222222222222222222222221
12222222222222222222222222222222222222222222222221
12222222222222222222222222222222222222222222222221
12222222222222222222222222222222222222222222222221
12222222222222222222222222222222222222222222222221
12222222222222222222222222222222222222222222222221
12222222222222222222222222222222222222222222222221
12222222222222222222222222222222222222222222222221
12222222222210122222222222222222222222222222222221
12222222222210122222222222222222222222222222222221
11111122221012222222222222222222222222222222222221
00002111110112222222222222222222222222222222222221
30002200000000222222222222222222222222222222222221
00000000000000222222222222222222222222222222222221
00000022011011222222222222222222222222222222222221
11110021111012222222222222222222222222222222222221
12211221221012222222222222222222222222222222222221
12222111222022222222222222222222222222222222222221
12222222222222222222222222222222222222222222222221
12222222222222222222222222222222222222222222222221
12222222222222222222222222222222222222222222222221
12222222222222222222222222222222222222222222222221
12222222222222222222222222222222222222222222222221
12222222222222222222222222222222222222222222222221
12222222222222222222222222222222222222222222222221
12222222222222222222222222222222222222222222222221
12222222222222222222222222222222222222222222222221
12222222222222222222222222222222222222222222222221
12222222222222222222222222222222222222222222222221
12222222222222222222222222222222222222222222222221
12222222222222222222222222222222222222222222222221
11111111111111111111111111111111111111111111111111
```

Figure 6.12: This is the frame used in the example dungeon encoding an entrance with a checkpoint and some forced corridor space at the center of the left side. The encoding is 0=empty, 1=rock, 2=mutable space to be filled later by the algorithm, and 3=a checkpoint which is also empty space.

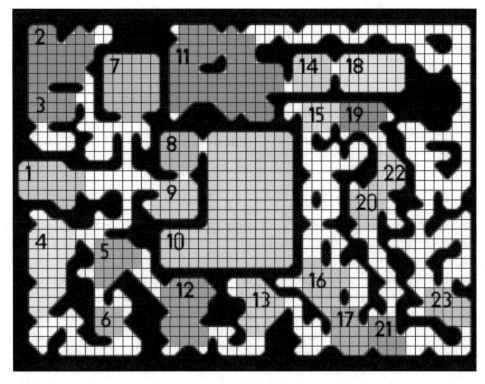

Figure 6.13: An example of an evolved level map. The **entrance** is room 1, the **goblin liar** fills rooms 8, 9, and 10. The **evil mage** is in rooms 14 and 18, the **armory** is in room 7.

3. Rooms 14 and 18, the lair of a powerful magical being, and

4. Room 12, an amorphous area designed to contain a trap.

The representation used to evolve levels is a type of large integer representation using integers in the range $0 \leq x < 100{,}000$. The gene consists of a linear array of 1728 large integers, more than are needed, which are used to fill a 48×36 rectangular array. The representation is generative and expressed using Algorithm 3.

The number 1728 is the number of squares in the dungeon. The term *slicing* an integer means using the mod and integer division operators to extract integers from the large integers making up the gene.

Algorithm 3 Expression algorithm

> The frame is loaded into a construction buffer.
> For each piece of required content
>   An integer is sliced to (x,y) in the buffer
>   Traversing the buffer in reading order,
>   content is placed in first open position at (x,y)
>   If not possible to place content, return fitness zero.
> End For
> Traversing the buffer in reading order
>   If a square is not yet filled, slice an integer to {0,1}
>   Fill the square with the sliced integer
> End Traverse

Once a dungeon is expressed we must compute its fitness. The entrances and all pieces of required content contain a square which is used as a checkpoint. A dynamic programming algorithm is then used to find the shortest path between all pairs of checkpoints. For each pair of checkpoints the user specifies if the algorithm wants the checkpoints to be far apart, close together, or that the user is indifferent to the distances between checkpoints. If any checkpoint cannot be reached from any other a fitness of zero is awarded, otherwise the fitness function is

$$fitness = \frac{\sum \text{long distances}}{1 + \sum \text{short distances}}. \tag{6.9}$$

This function is meant to be maximized. It is important to note that the values returned by this function are useful for relative ranking of members of a population but have no natural units of meaning. Two version of the fitness function are used in this test.

Definition 12 *The* **far fitness function** *specifies that the goblin lair is to be far from the entrance, the armory is to be near the goblin lair, and the magical being's lair is to be far from the entrance.*

Definition 13 *The* **near fitness function** *specifies that the goblin lair is to be near to the entrance, the armory is to be near the goblin lair, and the magical being's lair is to be far from the entrance.*

Figure 6.14 shows examples of three outcomes each of the map creation process for each of the two fitness functions. Alert readers will also notice that with five checkpoints there are ten pairs, and the exemplary fitness functions only use the value of three of these. While it is perfectly permissible to have many constraints, these constraints compete with one another, make the outcome unpredictable.

Evolutionary Algorithm Details

The evolutionary algorithm uses a population of 240 tentative dungeons. The algorithm is steady state [67] using single tournament selection of size seven. In a *mating event* a group of seven population members are chosen and the two best reproduce to replace the two worst. Reproduction

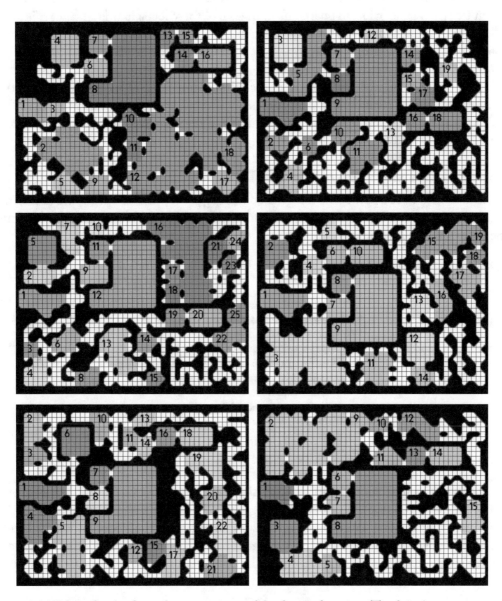

Figure 6.14: Maps for the far and near versions of the fitness function. The first three are examples of the far fitness function; the second three are examples of the near fitness function.

consists of first copying the parents over the structures to be replaced. The copies are then subjected to two-point crossover and mutation. The mutation operator first selects uniformly at random from 1–5 genetic loci to be modified. In each of these loci a new large integer is generated uniformly at random. Evolution is run for 100,000 mating events with summary statistics recorded every 1,000 mating events. The most fit final dungeon is then saved for additional processing.

Two experiments were run, one with each fitness function. In each experiment, 30 independent runs of the evolutionary algorithm are performed yielding 30 dungeon maps. These are all rendered and taken through steps (3) and (4) of the automatic FRPG module creation process.

6.3.3   IDENTIFYING AND CONNECTING ROOMS

A room is defined to be a part of the dungeon containing a 3×3 area of open space that forms a contiguous open space. The identification of contiguous open space is instantiated with a form of flood fill given in Algorithm 4.

Algorithm 4 Room membership algorithm

> **Mark the initial 3x3 area**
> **Set squares marked flag**
> **While(squares marked flag)**
>   **Reset squares marked flag**
>   **Forall empty squares**
>   **If (square has three marked neighbors)**
>   **Mark Squares**
>   **Set squares marked flag**
>   **End If**
>   **End For**
> **End While**

The room identification algorithm scans the map in transposed reading order looking for 3×3 open areas and, as each is located, calls Algorithm 4. This populates a "room membership" layer of the dungeon data structure. Open squares not assigned to a room are marked with a zero; other rooms are marked with their room numbers which form a set $1, 2, \ldots, n$ of rooms. Once the rooms of the dungeon are identified, the connectivity is abstracted using Algorithm 5.

Algorithm 5 Room adjacency algorithm

> **For each room**
> > **Copy the membership layer.**
> > **While empty rooms next to current room in copy.**
> > > **Change empty mark to current room mark.**
> > **End While**
> > **Scan copy**
> > > **Adjacency of a square of the extended current room**
> > > > **to any other room establishes adjacency.**
> > **End Scan**
> **End for**
> **Report adjacency relation.**

The room adjacency graph for the Example dungeon shown in Figure 6.13 is given in Figure 6.15. This example was taken from the runs using the *near* fitness function. Note that the constraints, goblin lair near the door, armory near the goblins, and evil mage far from the entrance are all clearly satisfied, based on the rendered graph. The graph was laid out with the *dot* software package [37]. Dot is an open graph specification and graph rendering language with a plethora of options.

While the map is the visually appealing version of the topology created by the evolutionary algorithm, it is worth noting that the graph is far smaller and contains the relevant connectivity information. This means that the algorithm that populates the adventure module may best be designed to operate on the graph representation of the level rather than the explicit map level.

6.3.4 POPULATING THE DUNGEON

Populating the dungeon consists of giving room descriptions which specify monsters, traps, treasures, and notes to the referee appropriate to each room. The populating algorithm uses the room adjacency graph, an example of which is shown in Figure 6.15, together with room size and saved checkpoints. The saved checkpoints permit the identification of the rooms with required content such as the entrance and goblin lair. Distance from rooms with checkpoints, e.g., distance from the entrance, also inform what is put in a room as does room size, another statistic available to the populating engine.

For each of the required content areas there are several alternate versions of the room descriptions which are selected uniformly at random. For other rooms there is a large selection of possible descriptions with a minimum size the room must be to use that content. Finally, each piece of text contains randomized objects such as **!R20-100silver!** which generates a random number of silver pieces in the range 20–100. In addition, there are magical object descriptors which contain random features like colors.

This method of populating the automatically created dungeon map permits the designer a good deal of control over the contents while varying the details so that players cannot get

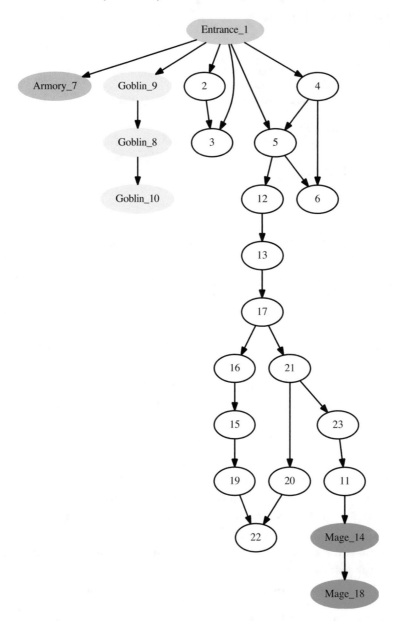

Figure 6.15: Shown is the adjacency graph for the rooms in the example level shown in Figure 6.13. The rooms associated with the entrance, goblins, armory, and magical being are colored.

those details from another group. To aid this effect, descriptions are written that have several similar rooms with different outcomes. A room with a trickle of water, for example, might have poisonous mold on the walls, treasure concealed behind a magical trap, or simply be a water source.

6.3.5   RESULTS FOR FRPG MODULE CREATION

Figure 6.14 shows six alternate dungeon maps. The first three are generated with a fitness function that rewards placing the goblin lair far from the entrance; the second three reward placing the goblin lair near the entrance. The form of the fitness function means that the two versions of the fitness function will have very different numerical ranges. This demonstrates that the fitness function, with its user-specified desirable properties, has value only for comparing maps within a single run.

The maps shown in Figure 6.14 show that putting the goblins far from the entrance is difficult; putting the mage's (far smaller) area far from the entrance is not so difficult. The six maps also show that the system yields a substantial diversity of outcomes.

The fitness functions used specify only a very small number of pairs of features with a desired near or far distance. During testing, a version of the fitness function that specified all pairs of distances was tested. This led to a phenomenon we term *transitive interference*, a particular type of over-constraining of the search problem. If A is to be near to B, B is to be far from C, and C is to be near to A then we create conflicting objectives that result from the transitive nature of distance constraints. When five features are used, as in the early testing performed, full specification yields ten objectives that are being fed into the single objective fitness function. The results are poor and, with the current form of the fitness function, only a small number of desired near or far distances should be given.

The use of a frame, introduced in this study, regularizes the map in convenient ways as intended. Note that in the map in Figure 6.13 and all six examples in Figure 6.14 room **1**, the entrance, is present at the same position. The use of the frame together with required features permits the game designer to use both rigid (frame) and floating (required content) objects in the automated design process. This enables modular, tiled design of the sort developed in [15] by putting connectors for tiles in the frame.

6.3.6   CONCLUSIONS AND NEXT STEPS FOR FRPG MODULE GENERATION

This study provides proof of concept for automatically generating playable FRPG modules. The process requires substantial user input in the form of a frame, required content for the map, and mutable text for both required content and rooms spontaneously generated by the level map evolver. Short of providing complete automation, the system presented here permits a designer to perform high level design and establish the flavor of the FRPG module. The advantage of this system is that it can generate dozens or hundreds of different versions of a given FRPG module all of which are faithful to the design elements specified by the user.

The different versions not only vary the details of the room descriptions but even the map and some of the treasures and monster encounters, while retaining those the designer deems critical. This sort of design methodology enables a higher degree of re-playability and can be structured to defeat the problem of players obtaining a copy of the dungeon module ahead of time.

6.3.7   DECORATIONS: MONSTERS, TREASURE, AND TRAPS

A topic left for future treatment is the provision of the less important decorations for the dungeon map. Beyond placing required content, like orcs or the big bad necromancer, this treatment neglects treasure and things like wandering monsters or causal encounters. This is largely because of available space and the topic is worthy of revisiting in the future. In general, treasure should appear behind monsters and traps should appear at navigation choke points, suggesting some simple, direct algorithmic approaches.

6.3.8   HISTORY, CONTEXT, AND STORY

A problem with procedurally generated content is that it can be generic; one version of the automatically generated adventure module can seem very much like another. An approach to avoiding this situation is to have backstory that make the adventure different. Examples might include the following.

- There was a flood in the recent history of the dungeon. Some of the rooms are knee-deep in water or mud, which can conceal traps or monsters like giant leeches. These decorations will help the dungeon be an individual rather than an instance of "not this again".

- A powerful mage brought down a comet on the location of the dungeon some time ago. This creates a well that punches through the top several levels of the dungeon, modifying the connectivity and lending a unique character to the maps.

- An earthquake collapsed one or more areas of the map. This can be done as a final processing step where the connectivity is reduced but not destroyed (or the need for a mage with a *move earth* spell or a squad of laborers with shovels is created).

When used in a commercial setting, automatic content generation is supposed to work well with the designers crafting a rich backstory. While required content can help with some of this—the comet strike is a geographic feature, for example—there is a great deal of room to augment the ideas in this chapter to create room for incorporating backstory elements.

The map creation software can also be augmented with *state variables*. Are the evil hordes that infest the map orcs, gnolls, zombies, or soldiers of the dark lord? Which of several wandering monster tables do you provide? Is there a preferred alignment for the creatures in the dungeon?

What sorts of treasure are available when decorating the dungeon? The identity of the big bad or a level boss can inform the settings that the content generator uses to create and populate the map. Other state variables can include temperature, moisture, decor, weather, and other mundane things that can give a map its own character.

Other than the adjacency and distance relations between the rooms and the size of rooms, the system reported in this chapter neither saves nor exploits state information. The use of state information by the generative code has the potential to make the module contain interesting puzzles. Examples of state information that might be incorporated include the following.

- State: magic weapons are present. Result: undead that may only be hit with magic weapons are available to the populating engine.

- State: a ring of fire protection is present. Result 1: walls of fire that block certain entrances or passages may be included by the populating engine. Result 2: treasures may be hidden in fires.

- State: one of the required content rooms is the lair of a necromancer. Result: zombies (dead animated by the necromancer) may be used by the populating engine.

Multicriteria Fitness

The current fitness function incorporates all of the designer's desires about near or far distances between checkpoints into a single fitness number. This study demonstrates that this is workable as long as the number of specifications by the designer remains small. Moving to multicriteria optimization will permit the system to deal far more effectively with designer requests. Parsing the final archive of non-dominated solutions will permit the designer to decide which of the constraints are more or less important by viewing level maps with poor or excellent values for those constraints. The incorporation of multicriteria optimization into this system is an early priority.

Doors

The maps used in this study have a total of no doors at all. In an FRPG module doors are an important type of object. They are a natural place to put traps and alarms; they conceal the content of the room behind them; and their construction and decoration is a natural place to establish the flavor of the module. Doors were relegated to future work in this study because they require substantial code to orient. The role of a door is very different on its two sides and so one of the clear next steps in this research is to add doors, including the ability to tell which side of the door is the inside and outside.

The dual maze design presented in [16] gives a way to make doors a part of level design. There are two maps, the one apparent to a creature like a giant spider who cannot open or close doors, and another apparent to creatures like the game players that can. Tactical control based on the ability to open doors has the potential to make an automatically designed level more interesting. Adding secret or concealed doors to the mix could add even more interest.

Bibliography

[1] A. Adamatzky, J. Serquera, and E. R. Miranda. *Automata: Theory and Applications of Cellular Automata: Cellular Automata Sound Synthesis: From Histograms to Spectrograms*. Luniver Press, 2008. 88

[2] A. Agapitos, J. Togelius, S. M. Lucas, J. Schmidhuber, and A. Konstantinidis. Generating diverse opponents with multiobjective evolution. In *Proc. of the Symposium on Computational Intelligence and Games*, pages 135–142, IEEE Press, Piscataway, NJ, 2008. DOI: 10.1109/cig.2008.5035632. 8

[3] B. Albertson. *Chess Mazes: A New Kind of Puzzle for Everyone*. South Chatham, MD, 2004. chesscafe.com 8

[4] B. Albertson. *Chess Mazes 2*. South Chatham, MD, 2008. chesscafe.com 8

[5] P. Anghelescu. Encryption algorithm using programmable cellular automata. *IEEE World Congress on Internet Security (WorldCIS)*, pages 233–239, 2011. 88

[6] D. Ashlock. *Evolutionary Computation for Optimization and Modeling*. Springer, New York, 2006. DOI: 10.1007/0-387-31909-3. 1, 111

[7] D. Ashlock. Automatic generation of game elements via evolution. In *Proc. of the Conference on Computational Intelligence in Games*, pages 289–296, IEEE Press, Piscataway, NJ, 2010. DOI: 10.1109/itw.2010.5593341. 8, 17, 28

[8] D. Ashlock. Evolvable fashion based cellular automata for generating cavern systems. In *Proc. of the Conference on Computational Intelligence in Games*, pages 306–313, IEEE Press, Piscataway, NJ, 2015. DOI: 10.1109/cig.2015.7317958. 98

[9] D. Ashlock, K. M. Bryden, and S. P. Gent. Evolutionary control of bracked L-system interpretation. In *Intelligent Engineering Systems Through Artificial Neural Networks*, vol. 14, pages 271–276, 2004. DOI: 10.1109/cec.2004.1331180. 45

[10] D. Ashlock, K. M. Bryden, and S. P. Gent. Creating spatially constrained virtual plants using L-systems. In *Smart Engineering System Design: Neural Networks, Evolutionary Programming, and Artificial Life*, pages 185–192, ASME Press, 2005. 45

[11] D. Ashlock, S. Gillis, A. McEachern, and J. Tsang. The do what's possible representation. In *Proc. of the IEEE Congress on Evolutionary Computation*, pages 1586–1593, 2016. DOI: 10.1109/cec.2016.7743978. 3

[12] D. Ashlock and B. Jamieson. Evolutionary computation to search Mandelbrot sets for aesthetic images. *Journal of Mathematics and Art*, 1(3), pages 147–158, 2008. DOI: 10.1080/17513470701585902. 117

[13] D. Ashlock, C. Lee, and C. McGuinness. Search based procedural generation of maze like levels. Submitted to the *IEEE Transactions on Computational Intelligence and Artificial Intelligence in Games*, 2010. DOI: 10.1109/tciaig.2011.2138707. 52, 56

[14] D. Ashlock, C. Lee, and C. McGuinness. Search-based procedural generation of maze-like levels. *IEEE Transactions on Computational Intelligence and AI in Games*, 3(3), pages 260–273, 2011. DOI: 10.1109/tciaig.2011.2138707. 17

[15] D. Ashlock, C. Lee, and C. McGuinness. Search-based procedural generation of maze-like levels. *IEEE Transactions on Computational Intelligence and AI in Games*, 3(3), pages 260–273, 2011. DOI: 10.1109/tciaig.2011.2138707. 74, 98, 129

[16] D. Ashlock, C. Lee, and C. McGuinness. Simultaneous dual level creation for games. *Computational Intelligence Magazine*, 2(6), pages 26–37, 2011. DOI: 10.1109/mci.2011.940622. 107, 131

[17] D. Ashlock, T. Manikas, and K. Ashenayi. Evolving a diverse collection of robot path planning problems. In *Proc. of the Congress on Evolutionary Computation*, pages 6728–6735, IEEE Press, Piscataway, NJ, 2006. DOI: 10.1109/cec.2006.1688530. 14, 22, 33

[18] D. Ashlock and C. McGuinness. Decomposing the level generation problem with tiles. In *Proc. of IEEE Congress on Evolutionary Computation*, pages 849–856, 2011. DOI: 10.1109/CEC.2011.5949707. 85, 107, 113

[19] D. Ashlock and C. McGuinness. Graph-based search for game design. *Game and Puzzle Design*, 2(2), pages 68–75, 2016. 5

[20] D. Ashlock and S. McNicholas. Fitness landscapes of evolved cellular automata. In Press, *IEEE Transaction on Evolutionary Computation*, 2013. DOI: 10.1109/tevc.2013.2243454. 86

[21] D. Ashlock and S. McNicholas. Fitness landscapes of evolved cellular automata. *IEEE Transaction on Evolutionary Computation*, 17(2), pages 198–212, 2013. DOI: 10.1109/tevc.2013.2243454. 89

[22] D. Ashlock and J. Montgomery. An adaptive generative representation for evolutionary computation. In *Proc. of the IEEE Congress on Evolutionary Computation*, pages 1578–1585, 2016. DOI: 10.1109/cec.2016.7743977. 3

[23] D. Ashlock, J. Schonfeld, L. Barlow, and C. Lee. Test problems and representations for graph evolution. In *Proc. of the IEEE Symposium on the Foundations of Computational Intelligence*, pages 38–45, 2014. DOI: 10.1109/foci.2014.7007805. 18

[24] D. Ashlock and M. Timmins. Adding local edge mobility to graph evolution. In *Proc. of the IEEE Congress on Evolutionary Computation*, pages 1594–1601, 2016. DOI: 10.1109/cec.2016.7743979. 3

[25] D. Ashlock and J. Tsang. Evolved art via control of cellular automata. In *IEEE Congress on Evolutionary Computation*, pages 3338–3344, 2009. DOI: 10.1109/cec.2009.4983368. 89

[26] D. Ashlock and A. Sherk. Non-local adaptation of artificial predators and prey. In *Proc. of the Congress on Evolutionary Computation*, vol. 1, pages 98–105, IEEE Press, 2005. DOI: 10.1109/cec.2005.1554665. 45, 74

[27] W. Ashlock. Using very small population sizes in genetic programming. In *Proc. of the Congress on Evolutionary Computation*, pages 319–326, 2006. DOI: 10.1109/cec.2006.1688325. 33

[28] W. Ashlock and D. Ashlock. Single parent genetic programming. In *Proc. of the Congress on Evolutionary Computation*, vol. 2, pages 1172–1179, IEEE Press, Piscataway, NJ, 2005. DOI: 10.1109/cec.2005.1554823. 86

[29] L. Barlow. *Representation, Graph Evolution, and the Induction of Desired Behaviours*. Ph.D. thesis, University of Guelph, 2015. 18

[30] R. Bellman. *Dynamic Programming*. Princeton University Press, Princeton, NJ, 1957. DOI: 10.1126/science.153.3731.34. 4

[31] C. Browne and F. Marie. Evolutionary game design. *IEEE Transactions on Computational Intelligence and AI in Games*, 2, pages 99–116, 2010. DOI: 10.1109/tciaig.2010.2041928. 8

[32] A. A. Burbelko, E. Fras, W. Kapturkiewicz, and D. Gurgul. Modelling of dendritic growth during unidirectional solidification by the method of cellular automata. *Materials Science Forum*, 649, pages 217–222, 2010. DOI: 10.4028/www.scientific.net/msf.649.217. 88

[33] A. A. Burbelko and D. Gurgul. Simulation of austenite and graphite growth in ductile iron by means of cellular automata. *Archives of Metallurgy and Materials*, 55(1), pages 53–60, 2010. 88

[34] E. V. Denardo. *Dynamic Programming: Models and Applications*. Prentice Hall, Englewood Cliffs, NJ, 1984. 4

[35] M. Devetakovic, L. Petrusevski, M. Dabic, and B. Mitrovic. Les folies cellulaires: An exploration in architectural design using cellular automata. *12th Generative Art Conference*, pages 181–192, 2009. 89

[36] G. B. Ermentrout and L. Edelstein-Keshet. Cellular automata approaches to biological modeling. *Journal of Theoretical Biology*, 160(1), pages 97–133, 1993. DOI: 10.1006/jtbi.1993.1007. 88

[37] E. R. Gansner and S. C. North. An open graph visualization system and its applications to software engineering. *Software—Practice and Experience*, 30(11), pages 1203–1233, 2000. DOI: 10.1002/1097-024x(200009)30:11%3C1203::aid-spe338%3E3.3.co;2-e. 127

[38] F. Gruau and L. D. Whitley. Adding learning to the cellular development of neural networks: Evolution and the Baldwin effect. *Evolutionary Computation*, pages 213–233, 1993. DOI: 10.1162/evco.1993.1.3.213. 18

[39] P. F. Hingston, L. C. Barone, and Z. Michalewics. *Design by Evolution*. Natural Computing Series, Springer, New York, 2008. DOI: 10.1007/978-3-540-74111-4. 18

[40] T. W. Hungerford. *Algebra*. Springer, 1974. DOI: 10.1007/978-1-4612-6101-8. 112

[41] R. Hunicke and V. Chapman. AI for dynamic difficulty adjustment in games. In *Proc. of the Workshop on Challenges in Game Artificial Intelligence*, pages 91–96, AAAI, 2004. 8

[42] L. Johnson, G. N. Yannakakis, and J. Togelius. Cellular automata for real-time generation of infinite cave levels. In *Proc. of the Workshop on Procedural Content Generation in Games*, *PCGames'10*, New York, ACM, 2010. DOI: 10.1145/1814256.1814266. 1

[43] L. Johnson, G. N. Yannakakis, and J. Togelius. Cellular automata for real-time generation of infinite cave levels. In *Proc. of the Workshop on Procedural Content Generation in Games*, *PCGames'10*, pages 10:1–10:4, ACM, New York, 2010. DOI: 10.1145/1814256.1814266. 87, 92

[44] P. Cardwell Jr. The attacks on role-playing games. *Skeptical Inquirer*, 18(2), pages 157–165, 1994. 120

[45] R. Kicinger, T. Arciszewski, and K. De Jong. Evolutionary computation and structural design: A survey of the state-of-the-art. *Computers and Structures*, pages 1943–1978, 2005. DOI: 10.1016/j.compstruc.2005.03.002. 18

[46] D. E. Knuth. *The Art of Computer Programming Volume 2: Seminumerical Algorithms*. Addison-Wesley, New York, 1997. 5

[47] A. Lindenmayer. Mathematical models for cellular interaction in development, parts I and II. *Journal of Theoretical Biology*, 16, pages 280–315, 1968. DOI: 10.1016/0022-5193(68)90080-5. 45, 66

[48] M. E. L'raga and L. Alvarez-Icaza. Cellular automaton model for traffic flow based on safe driving policies and human reactions. *Physica A*, 389(23), pages 5425–5438, 2010. DOI: 10.1016/j.physa.2010.08.020. 88

[49] J. Marks and V. Hom. Automatic design of balanced board games. In *Proc. of the Artificial Intelligence and Interactive Digital Entertainment International Conference (AIIDE)*, pages 25–30, AAAI, 2007. 8

[50] C. McGuinness and D. Ashlock. Incorporating required structure into tiles. In *Proc. of CIG*, pages 16–23, 2011. DOI: 10.1109/cig.2011.6031984. 85, 113

[51] T. Mikkel. Undirected single-source shortest paths with positive integer weights in linear time. *Journal of the ACM*, 3(46), pages 362–394, 1999. DOI: 10.1145/316542.316548. 5

[52] G. Monro. Emergence and generative art. *Leonardo*, 42(5), pages 476–477, MIT Press, 2009. DOI: 10.1162/leon.2009.42.5.476. 89

[53] J. Montgomery and D. Ashlock. Applying the biased form of the adaptive generative representation. In *Proc. of the Congress on Evolutionary Computation*, pages 1079–1086, IEEE Press, Piscataway, NJ, 2017. DOI: 10.1109/cec.2017.7969427. 19

[54] K. Nakamura and K. Imada. Incremental learning of cellular automata for parallel recognition of formal languages. In *Proc. of the 13th International Conference on Discovery Science, DS'10*, pages 117–131, Springer-Verlag, Berlin, Heidelberg, 2010. DOI: 10.1007/978-3-642-16184-1_9. 88

[55] S. B. Needleman and C. D. Wunsch. A general method applicable to the search for similarities in the amino acid sequence of two proteins. *Journal of Molecular Biology*, 48(3), pages 443–453, 1970. DOI: 10.1016/0022-2836(70)90057-4. 4

[56] Y. I. H. Parish and P. Müller. Procedural modeling of cities. In *Proc. of the 28th Annual Conference on Computer Graphics and Interactive Techniques, SIGGRAPH'01*, pages 301–308, ACM, New York, 2001. DOI: 10.1145/383259.383292. 45

[57] D. Perez, E. J. Powley, D. Whitehouse, P. Rohlfshagen, S. Samothrakis, P.I. Cowling, and S. M. Lucas. Solving the physical traveling salesman problem: Tree search and macro actions. *IEEE Transaction on Computational Intelligence and AI in Games*, 6(1), pages 31–45, 2014. DOI: 10.1109/tciaig.2013.2263884. 5

[58] P. Prusinkiewicz, A. Lindenmayer, and J. S. Hanan. *The Algorithmic Beauty of Plants*. Springer-Verlag, New York, 1990. DOI: 10.1007/978-1-4613-8476-2. 45, 66

[59] F. Rothlauf. Representations for evolutionary algorithms. In *Proc. of the GECCO Conference Companion on Genetic and Evolutionary Computation*, pages 2613–2638, New York, ACM, 2008. DOI: 10.1145/3067695.3067718. 17

[60] E. Sapin, O. Bailleux, and J. Chabrier. Research of complexity in cellular automata through evolutionary algorithms. *Complex Systems*, 11, 1997. 88

[61] J. Serquera and E. R. Miranda. Cellular automata sound synthesis with an extended version of the multitype voter model. In *Audio Engineering Society Convention 128*, 5, 2010. 88

[62] J. Serquera and E. R. Miranda. *Applications of Evolutionary Computation: Evolutionary Sound Synthesis: Rendering Spectrograms from Cellular Automata Histograms*. Springer, Berlin/Heidelberg, 2010. DOI: 10.1007/978-3-642-12242-2_39. 89

[63] V. Singh and N. Gu. Towards an integrated generative design framework. In Press, *Design Studies*, 2011. DOI: 10.1016/j.destud.2011.06.001. 89

[64] H. Situngkir. Exploring ancient architectural designs with cellular automata. *BFI Working Paper No. WP-9-2010*, 2010. DOI: 10.2139/ssrn.1696683. 89

[65] N. Sorenson and P. Pasquier. Towards a generic framework for automated video game level creation. In *Proc. of the European Conference on Applications of Evolutionary Computation (EvoApplications)*, vol. 6024, pages 130–139, Springer LNCS, 2010. DOI: 10.1007/978-3-642-12239-2_14. 1

[66] G. J. Sullivan and R. L. Baker. Efficient quadtree coding of images and video. *IEEE Transactions on Image Processing*, 3(3), 1994. DOI: 10.1109/83.287030. 75

[67] G. Syswerda. A study of reproduction in generational and steady state genetic algorithms. In *Foundations of Genetic Algorithms*, pages 94–101, Morgan Kaufmann, 1991. DOI: 10.1016/b978-0-08-050684-5.50009-4. 29, 76, 90, 124

[68] J. Togelius and J. Schmidhuber. An experiment in automatic game design. In *Proc. of the IEEE Symposium on Computational Intelligence and Games*, pages 111–118, AAAI, 2008. DOI: 10.1109/cig.2008.5035629. 8

[69] J. Togelius, G. Yannakakis, K. Stanley, and C. Browne. Search-based procedural content generation. In *Applications of Evolutionary Computation*, v. 6024 of *Lecture Notes in Computer Science*, pages 141–150, Springer, Berlin/Heidelberg, 2010. DOI: 10.1007/978-3-642-12239-2_15. 1

[70] J. Togelius, M. Preuss, and G. N. Yannakakis. Towards multiobjective procedural map generation. In *Proc. of the Workshop on Procedural Content Generation in Games, PCGames'10*, pages 1–8, New York, ACM, 2010. DOI: 10.1145/1814256.1814259. 1

[71] A. J. Viterbi. Error bounds for convolutional codes and an asymptotically optimum decoding algorithm. *IEEE Transactions on Information Theory*, 13(2), 1967. DOI: 10.1109/tit.1967.1054010. 4

[72] L. D. Whitley and L. D. Pyeatt F. Gruau. Cellular encoding applied to neurocontrol. In *Proc. of ICGA*, pages 460–469, 1995. 18

[73] S. Wolfram. Universality and complexity in cellular automata. *Physica D: Nonlinear Phenomena*, 10(1–2), pages 1–35, 1984. DOI: 10.1016/0167-2789(84)90245-8. 88

Author's Biography

DANIEL ASHLOCK

Dr. Daniel Ashlock is a professor of mathematics at the University of Guelph in Guelph, Ontario, Canada. He is a member of the IEEE Computational Intelligence Society Games Technical Committee, serves as an associate editor of both the *IEEE Transactions on Games* and the new journal *Game and Puzzle Design*. Dr. Ashlock is a lifelong referee of role-playing games and loves the problems that arise in game and puzzle design. He is the author of 270 peer-reviewed scientific publications, about one-third of which are on the topic of games. Across his work the issue of representation, the study of how to phrase problems for the computer, appears in a starring role. Dr. Ashlock is the Chief of Innovation at Ashlock and McGuinness Consulting, Inc.

Printed in the United States
by Baker & Taylor Publisher Services